Social Security Administration Electronic Service Provision

A STRATEGIC ASSESSMENT

Leon J. Osterweil, Lynette I. Millett, and Joan D. Winston, *Editors*

Committee on the Social Security Administration's E-Government Strategy and Planning for the Future

Computer Science and Telecommunications Board

Division on Engineering and Physical Sciences

NATIONAL RESEARCH COUNCIL
OF THE NATIONAL ACADEMIES

THE NATIONAL ACADEMIES PRESS
Washington, D.C.
www.nap.edu

THE NATIONAL ACADEMIES PRESS 500 Fifth Street, N.W. Washington, DC 20001

NOTICE: The project that is the subject of this report was approved by the Governing Board of the National Research Council, whose members are drawn from the councils of the National Academy of Sciences, the National Academy of Engineering, and the Institute of Medicine. The members of the committee responsible for the report were chosen for their special competences and with regard for appropriate balance.

Support for this project was provided by the Social Security Administration with assistance from the National Science Foundation under Grant No. CNS-0344585. Any opinions, findings, conclusions, or recommendations expressed in this publication are those of the authors and do not necessarily reflect the views of the organizations or agencies that provided support for the project.

International Standard Book Number-13: 978-0-309-10393-0
International Standard Book Number-10: 0-309-10393-2

Additional copies of this report are available from:

The National Academies Press
500 Fifth Street, N.W., Lockbox 285
Washington, DC 20055
(800) 624-6242
(202) 334-3313 (in the Washington metropolitan area)
http://www.nap.edu

Printed in the United States of America

THE NATIONAL ACADEMIES

Advisers to the Nation on Science, Engineering, and Medicine

The **National Academy of Sciences** is a private, nonprofit, self-perpetuating society of distinguished scholars engaged in scientific and engineering research, dedicated to the furtherance of science and technology and to their use for the general welfare. Upon the authority of the charter granted to it by the Congress in 1863, the Academy has a mandate that requires it to advise the federal government on scientific and technical matters. Dr. Ralph J. Cicerone is president of the National Academy of Sciences.

The **National Academy of Engineering** was established in 1964, under the charter of the National Academy of Sciences, as a parallel organization of outstanding engineers. It is autonomous in its administration and in the selection of its members, sharing with the National Academy of Sciences the responsibility for advising the federal government. The National Academy of Engineering also sponsors engineering programs aimed at meeting national needs, encourages education and research, and recognizes the superior achievements of engineers. Dr. Charles M. Vest is president of the National Academy of Engineering.

The **Institute of Medicine** was established in 1970 by the National Academy of Sciences to secure the services of eminent members of appropriate professions in the examination of policy matters pertaining to the health of the public. The Institute acts under the responsibility given to the National Academy of Sciences by its congressional charter to be an adviser to the federal government and, upon its own initiative, to identify issues of medical care, research, and education. Dr. Harvey V. Fineberg is president of the Institute of Medicine.

The **National Research Council** was organized by the National Academy of Sciences in 1916 to associate the broad community of science and technology with the Academy's purposes of furthering knowledge and advising the federal government. Functioning in accordance with general policies determined by the Academy, the Council has become the principal operating agency of both the National Academy of Sciences and the National Academy of Engineering in providing services to the government, the public, and the scientific and engineering communities. The Council is administered jointly by both Academies and the Institute of Medicine. Dr. Ralph J. Cicerone and Dr. Charles M. Vest are chair and vice chair, respectively, of the National Research Council.

www.national-academies.org

COMMITTEE ON THE SOCIAL SECURITY ADMINISTRATION'S E-GOVERNMENT STRATEGY AND PLANNING FOR THE FUTURE

LEON J. OSTERWEIL, University of Massachusetts, Amherst, *Chair*
MATT BISHOP, University of California, Davis
MICHAEL J. CAREY, BEA Systems, Inc.
DAVID J. DeWITT, University of Wisconsin-Madison
VALERIE GREGG, University of Southern California, Information Sciences Institute
BLAISE HELTAI, New Vantage Partners
STEPHEN H. HOLDEN, Touchstone Consulting Group
LARRY G. MASSANARI, Social Security Administration (retired)
JUDITH S. OLSON, University of Michigan

Staff

LYNETTE I. MILLETT, Study Director and Senior Program Officer
JOAN D. WINSTON, Program Officer
JANICE M. SABUDA, Senior Program Assistant

For more information on CSTB, see its Web site at http://www.cstb.org, write to CSTB, National Research Council, 500 Fifth Street, N.W., Washington, DC 20001, call (202) 334-2605, or e-mail CSTB at cstb@nas.edu.

Preface

The use of the Internet (and other information technology) among the general population has resulted in a rising level of comfort and familiarity with these technologies. Businesses such as retailers, banks, and investment companies have been shifting more of their operations online, seeking to meet customer demand while reducing costs. Many businesses offer customer services online (these electronic services are often called e-business), often 24 hours a day, 7 days a week. The public sector has also embraced these technologies, although their adoption in the public sector often lags that of the private sector's more aggressive e-business adopters. Much like the case in the private sector, governments' basic goals for such automation include satisfying customer service expectations and increasing the efficiency and effectiveness of operations.

The Social Security Administration (SSA), a federal agency that interacts with broad segments of the public, has been developing online government services for more than a decade. The SSA's e-government initiatives thus far have included support of online transactions relating to benefits applications, requests for statements, replacement Medicare cards, and disability reports.

The SSA's clients include not only nearly all U.S. residents (both those contributing as workers and those collecting benefits) but also millions of employers filing wage reports. In addition to providing direct services to citizens and employers, agencies such as the SSA are mandated to coordinate and cooperate in various ways (for example, through data exchange or service provision) with other state and federal agencies.

Like the underlying technologies, innovation and deployment of electronic services in the commercial sector continue at a rapid pace. As this report describes, this pace exerts pressure on federal agencies to improve their services continually and to stay abreast of both technology developments and the associated developments in business practices and technology management. Perhaps more importantly, broader deployment and adoption of electronic services offer agencies like the SSA potential relief from the increasing workload, workforce, and other resource pressures facing them.

To further understanding of these and related issues, the SSA's Deputy Associate Commissioner for Electronic Services asked the National Research Council (NRC) to examine the SSA's proposed e-government strategy and the underlying service-delivery and information technology infrastructure and to prepare a report discussing issues including the following: the SSA's current e-government strategy, including technological assumptions, performance measures and targets, planned operational capabilities, strategic requirements, and future goals; strategies, assumptions, and technical and operational requirements in comparable public- and private-sector institutions, and their implications for the SSA; and ongoing efforts to define and refine the SSA's long-term strategy.

The Committee on the Social Security Administration's E-Government Strategy and Planning for the Future was appointed under the auspices of the NRC's Computer Science and Telecommunications Board to conduct the study. The nine members of the study committee have expertise in areas such as software engineering and methodology, e-business, e-government, information system security, databases, data and application integration, application of technology to business transformation, project management and decision-support systems, human-computer interaction, and SSA operations and management. Biographical information for members of the committee and the NRC staff is presented in Appendix A. (Ken Orr of the Ken Orr Institute resigned from the committee in September 2005 owing to time constraints.)

The committee held three meetings during the course of its work. One was an organizational meeting by teleconference and two featured testimony (1) to gather information on the SSA's current e-government strategy, including technical requirements and strategies of the agency, and to learn what the agency believes are roadblocks to potential success, as well as what positive outcomes are anticipated; and (2) to gather related information from representatives of some of the SSA's key constituencies and to discuss comparable systems issues in other institutions (in both the public and private sectors). Panelists and briefers for the meetings are listed in Appendix B.

In late spring of 2007, during a final fact-checking phase in the course of preparing this report, the committee requested an update from the SSA regarding any additional steps taken in consideration of converting its Master Data Access Method (MADAM) database system and any changes to the SSA's organizational structure affecting electronic services. The committee was informed that there were no substantive updates.

The committee focused on examining the SSA's current e-government strategy, including technological assumptions, operational capabilities, functional requirements, and future goals and ongoing efforts to define and refine the SSA's long-term strategy to support information technology applications and online services to its many and varied constituencies. Consistent with early discussions with and briefings from the SSA, this report assumes that the SSA intends to, and should, pursue delivering an expanding array of online services. Questions of whether online services are appropriate for the SSA or what the balance of resources devoted to online services and other modes of delivery should be were considered to be beyond the scope of this study. In keeping with the resources available for the study, the report does not undertake to develop a comprehensive roadmap to take the agency from its current situation to a future involving more online services. Finally, although the report acknowledges the importance of privacy to individuals and to society, it does not provide a comprehensive examination of the agency's privacy policies and safeguards.

The committee thanks the many individuals who contributed to its work. It appreciates the panelists' and the SSA's willingness to address the questions posed to them and is grateful for their insights. The study's sponsors at the Social Security Administration and the SSA staff have been most supportive and responsive in helping the committee to do its work. We further wish to recognize the energetic participation of the SSA meeting attendees as a group. The reviewers of this report provided constructive feedback and insights, and we are grateful for their assistance. The committee would also like to express its thanks to the members of the staff of the National Academies, especially to study director Lynette Millett and program officer Joan Winston, who displayed exemplary professionalism and patience in seeing this challenging project through to a satisfying conclusion, and to Janice Sabuda, who facilitated our meeting and other activities through the course of the project.

Leon J. Osterweil, *Chair*
Committee on the Social Security
Administration's E-Government Strategy and
Planning for the Future

Acknowledgment of Reviewers

This report has been reviewed in draft form by individuals chosen for their diverse perspectives and technical expertise, in accordance with procedures approved by the National Research Council's (NRC's) Report Review Committee. The purpose of this independent review is to provide candid and critical comments that will assist the institution in making its published report as sound as possible and to ensure that the report meets institutional standards for objectivity, evidence, and responsiveness to the study charge. The review comments and draft manuscript remain confidential to protect the integrity of the deliberative process. We wish to thank the following individuals for their review of this report:

Michael Brodie, Verizon Communications,
Sharon Dawes, State University of New York at Albany,
Carlo De Luca, Boston University,
Marc Donner, Morgan Stanley,
Sara Kiesler, Carnegie Mellon University,
Darrell Long, University of California, Santa Cruz,
Ken Nibali, Independent Consultant, West Friendship, Maryland,
Daniel Schutzer, Financial Services Technology Consortium,
Peter Weinberger, Google Inc., and
Marsha Young, Booz Allen Hamilton.

Although the reviewers listed above have provided many constructive comments and suggestions, they were not asked to endorse the conclusions or recommendations, nor did they see the final draft of the report before its release. The review of this report was overseen by William H. Press, Los Alamos National Laboratory. Appointed by the National Research Council, he was responsible for making certain that an independent examination of this report was carried out in accordance with institutional procedures and that all review comments were carefully considered. Responsibility for the final content of this report rests entirely with the authoring committee and the institution.

Contents

Summary

This report examines the Social Security Administration's (SSA's) proposed e-government strategy and provides advice on how the SSA can best deliver services to its constituencies in the future. The assessment by the Committee on the Social Security Administration's E-Government Strategy and Planning for the Future was based on (1) its examination of the SSA's current e-government strategy, including technological assumptions, performance measures and targets, planned operational capabilities, strategic requirements, and future goals; (2) its consideration of strategies, assumptions, and technical and operational requirements in comparable public- and private-sector institutions; and (3) its consideration of the larger organizational, societal, and technological context in which the SSA operates.

CHALLENGES AND OPPORTUNITIES FOR THE SOCIAL SECURITY ADMINISTRATION

The SSA's operations are extensive—one or more of its programs touch the lives of most Americans, from the almost 160 million workers who pay Social Security taxes on wages, to the nearly 50 million people who receive benefits under the Old Age and Survivors Insurance program, to the roughly 8 million disabled individuals and eligible family members who receive benefits under the Disability Insurance program. The SSA also administers the Supplemental Security Income (SSI) program and makes needs-based determinations of eligibility for payments under the program.

The SSA faces demographic challenges and growing public expectations as it conducts a broad scope of activities, services, and interactions. It is predicted that the SSA's workload will increase sharply as the baby boomers—the large cohort born during the 1946-1964 period, sometimes called the Silver Tsunami—reach retirement or become disability-prone. For example, the number of people filing for retirement annually has increased by 500,000 since 2000, a 25 percent increase.[1] At the same time, like other federal agencies with an aging workforce, the SSA is facing a projected brain drain. A substantial number of the agency's most experienced employees (who best understand the complex benefits-determination processes and the agency's large and complex technology infrastructure) could retire at any time.

Today, public contact with the SSA largely takes place face to face at its field offices, by phone through the teleservice centers, or through the mail (for example, Social Security statements are automatically mailed yearly to workers over the age of 25). These activities are labor-intensive; over 60 percent of the SSA's employees (located mainly in field offices and at teleservice centers) deliver direct service to the public, and another 30 percent (in the regional offices, processing centers, and headquarters) provide direct support to those front-line workers.

Some sectors of the economy have seen a broad push toward online services as both a complement to and a substitute for traditional service-delivery mechanisms. For reasons including cost-effectiveness and enhanced customer satisfaction, many commercial organizations today are offering online, e-business services, often 24 hours a day, 7 days a week (24/7), 365 days a year. The growing adoption and use of such services reflect the emergence of a suite of information technologies capable of supporting online services and a general public that has a rising level of comfort and familiarity with the Internet and other information technologies in personal, social, and commercial contexts and that increasingly expects both firms and government agencies to provide online information and services.

As they do for commercial enterprises, online services offer the SSA an opportunity to improve its operational efficiency and to increase its total service capacity—in particular, to cope with its growing workload at a time when it is facing its own retirement wave. The technologies for providing online services have reached a sufficient level of maturity to

[1]Mary Mosquera, "Case Files Travel Lighter, Faster," *Government Computer News*, Oct. 9, 2006, available at http://www.gcn.com/print/25_30/42177-1.html, accessed June 14, 2007.

enable the realization of this goal. Additionally, many in the new wave of retirees are likely to be computer-literate and Internet-savvy (and to retain these traits after reaching retirement age); they will thus be more adept and frequent users of online services than were previous generations of SSA beneficiaries. This does not mean that the SSA should go to all-electronic service, but it does mean that a larger proportion of its clients and beneficiaries will be able to use, and may even prefer, online service channels—paths through which they can accomplish their goals—to the extent possible.

Indeed, the SSA uses a range of online services as a complement to the traditional service-delivery channels, and it looks to information technology (IT) to help provide services to its varied user communities: clients (current and prospective SSA beneficiaries), the general public, and partners such as the states and other federal agencies, along with various types of third parties (such as third-party representatives, representative payees, employers that pay workers' payroll taxes, and so on). However, the SSA currently does not strongly promote online services as an alternative to traditional service-delivery channels. In addition, with respect to the SSA's e-government initiatives, the focus of this report, responsibilities for software, hardware, and support for providing electronic information and services are split across the domains of several SSA Deputy Commissioners.

Finding: The SSA's organizational structure does not support the establishment of a strategic focus in electronic services that is sufficiently high-level and broad-based. The SSA has an opportunity to be more proactive in fundamentally reassessing its customer service value chain and, for as many customers as possible, focusing on the potential substitution of electronic services for other delivery channels, such as paper mail and face-to-face interactions in field offices.

Recommendation: The SSA should make an unambiguous, strategic commitment to electronic services as part of its long-term service-delivery strategy, placing a central emphasis on electronic services that encompass timely and up-to-date information for users, partners, and beneficiaries.

LEARNING FROM THE EXPERIENCES OF LARGE FINANCIAL INSTITUTIONS

Although the SSA is in some ways a distinctive institution, in many respects its activities, information technology, and other operational characteristics resemble those of large financial institutions (banks and brokerages). These institutions maintain a very large number of accounts and receive and make a high volume of payments. They usually conduct transactions in multiple customer segments (individual retail customers, businesses, other financial institutions, and so on) and through multiple service channels (online, through automated teller machines and branch offices, through call centers, and so on). The analogy is also apt in terms of IT requirements; the databases maintained by the largest of the private-sector institutions are now quite comparable, both in size and in transaction volume, to the databases for which the SSA is responsible.

The financial services community has been one of the most aggressive and competitive in using IT and electronic commerce, and banks and brokerages have made significant investments in developing their online service channels. The business case for doing so is composed of multiple elements: customer satisfaction, customer retention, new-customer acquisition, cross-sales of other products to existing customers, and the economic trade-offs between cost reduction (including savings from reduced error rates and from not needing to redo work to correct errors) and the comparative cost of delivering online services. In particular, the cost per transaction differs greatly among service-delivery channels, by some accounts by an order of magnitude.

Because of these advantages, banks aggressively market online services to attract customers to these channels and to retain customers in them. Experience has shown that the successful introduction of electronic services leads to customers' use of other available electronic services, thus compounding the benefits of the investment in such services. A satisfied user of one online service is more likely to seek out similar services in order to avoid long telephone queues or paper-based cycle times.

Importantly, even as they have greatly increased their emphasis on online operations, banks have not abandoned their branch offices or call centers. Rather, they consider online services as one important aspect of providing services and continually balance and rebalance their portfolio of service offerings. Indeed, banks still make heavy use of call centers for servicing customers—although the call centers themselves have been transformed. With the convergence of voice and data (Voice over Internet Protocol, or VoIP) and the use of speech recognition, these call centers are becoming very heavily automated and very cost-effective. Banks are exploring in which situations to have self-service and in which it is desirable to have human agents, as well as experimenting with collaborative Web sites that can draw on assistance from human agents. For example, for simple transactions and information access, self-service is offered; for more complex transactions and cross-sell opportunities, more direct interaction with humans is used. Analogously, the committee does not suggest that the SSA should offer online services exclusively, but that it should be prepared to balance and rebalance its service channels in order to meet its efficiency and effectiveness goals and the expectations of its public.

The SSA might not be able to—or want to—use "as is" all of the approaches and solutions pursued by other institutions. The committee believes that there is nonetheless substantial value to the SSA in considering what might be learned and what might be adapted or applied as a result of studying appropriate analogous organizations—of which financial institutions are a leading example. Although the SSA's operations, customer base, and transaction patterns are not identical with those of a large financial institution, the committee believes that examination of the commercial financial services industry's experience, market research, and product-refinement knowledge would be useful to the SSA and to other government agencies seeking to make more comprehensive and effective transitions to online services.

Finding: The experiences of large-scale financial institutions in transitioning to the provision of electronic services are instructive in considering the challenges faced by the SSA in formulating its medium- and long-term electronic services strategy.

Recommendation: The SSA should carefully consider the ways in which the experiences and approaches of large-scale financial institutions—including state-of-the-practice electronic information and service delivery, metrics-guided improvement, and process transformation, among other approaches and solutions—might be relevant to the kinds of services that the agency is providing or may provide in the future.

The strategic and management approaches that financial institutions have adopted to launch and develop online services are also especially instructive. In the largest and most successful financial companies, commercial management and organizational structures for online services have gone through three phases:

- In the first phase, which for most banks was 1996 to 1998, multiple "e-groups" were emerging in the organization. The primary focus was often in marketing and communications organizations, as the chief use of Web sites was to market services rather than to provide them.
- In the second phase (approximately 1999 to 2002) "e-business" was paramount. Centralized, autonomous e-commerce lines of business—reporting very high in the organization—were formed with virtually end-to-end control of the e-channel. This seems, in retrospect, an almost necessary step to developing a sophisticated strategy, infrastructure, and set of policies for electronic service provision.

- In the current state of maturity (approximately 2002 to the present), the motto appears to be "e-business is business," and electronic services have become reabsorbed as integral parts of the organization's lines of business.

Note that this progression, which the committee believes closely mirrors the processes that the SSA will need to follow, is characteristic of institutions that have had to undergo significant transformation to take advantage of electronic service provision. (Note also that the strategy and evolution of companies that were Internet-based from the outset, such as Amazon or eBay, hold fewer organizational lessons for the SSA because they have not had to revamp their technology, operations, and organization.)

The SSA's management and organizational structures for electronic services and e-government have not yet moved to the second phase of electronic services maturity. When moving from the first to the second phase in this progression, the large-scale financial institutions referred to above established centralized focal points for electronic services. These focal points typically reported very high in the organization. The committee believes that the SSA should follow a trajectory similar to that described above and that the development and management of electronic services should be centralized and elevated in the organization.

Finding: The SSA's present direction diverges from the three-phase progression that large financial institutions have followed in successfully developing and launching electronic services.

Recommendation: In order to move to the second phase of electronic services maturity, the SSA should create a focal point responsible for developing and managing electronic information and service delivery—including components such as Web content, online transactions, user interfaces, research, database systems and other key enabling technologies, and other facets of electronic service delivery that are currently dispersed throughout the SSA. This focal point should have sufficient resources to take on organization-wide responsibility for online services and should report directly to the SSA Commissioner or to a Deputy Commissioner.

TECHNOLOGICAL LEGACY POSES RISKS AND HINDERS DEVELOPMENT OF ELECTRONIC SERVICES

The SSA's Master Data Access Method (MADAM) is a database system that was developed in-house by SSA staff in the early 1980s when the SSA converted from tape to disk storage for its data sets. Although MADAM is still in use today, this system is technologically obsolete and functionally primitive compared with readily available commercial technologies and products; its current status has several important implications.

First, this technology will constrain the SSA as it continues to develop its online services. Relying on MADAM means that the SSA's systems are not able to exploit modern database access tools and utilities that facilitate support for the full suite of electronic services provided by the private sector. Also, whereas contemporary commercial products support near-24/7 access, information provided to the committee indicates that updates to the central MADAM database are done primarily (if not exclusively) through periodic, lengthy batch updates, which make 24/7 access impossible. Thus, MADAM could be an obstacle to the kind of user acceptance needed to reduce the SSA's costs of transactions with its growing user community.

Second, although MADAM continues to function, it exposes the SSA to a number of significant risks. The use of MADAM requires that the SSA and its contractors maintain an obsolete, custom system that is built on top of an increasingly antiquated underlying technology base, in which some of the software is written in the Common Business-Oriented Language, or COBOL. Maintaining this system requires very specialized expertise that is becoming increasingly scarce. It is also unclear how easy it will be to continue to find expertise in more-generic but increasingly obsolete software technologies such as the COBOL programming language. Related to this, the underlying programming technology of the SSA's core systems makes these systems more cumbersome to maintain compared with those implemented using modern technology. Moreover, the continued use of MADAM (and other now-antiquated similar solutions such as Computer Associates Integrated Database Management System [CA-IDMS]) locks the SSA into a single, expensive hardware and software combination, which precludes the agency from taking advantage of potentially lower cost alternatives.

In 1986, a report from the congressional Office of Technology Assessment (OTA)[2] alerted the SSA that the technology and technological risks inherent in MADAM make it a serious liability. However, MADAM is still in use. One argument that has been offered for retaining MADAM is that the SSA's databases are so large that they can only be handled by custom software. Today, however, other enterprises have databases that are com-

[2]U.S. Congress, Office of Technology Assessment, *The Social Security Administration and Information Technology*, OTA-CIT-311, Washington, D.C.: U.S. Government Printing Office, 1986, p. 43 (NTIS Order PB87-136834), also available at http://www.ssa.gov/history/pdf/ota86.pdf, accessed June 20, 2007.

parable in size to those of the SSA, and commercial database products are up to the task. Another likely reason that the SSA has not moved to replace MADAM is that the migration to modern database technology would be far from straightforward. Indeed, the committee concurs that a migration will be difficult, but it believes that failing to migrate also poses considerable risks and constrains future delivery of online services.

The SSA has been exploring a particular path—to convert the MADAM-resident data sets to a commercially available system, IBM's relational database system product, DB2, but to avoid rewriting existing applications by changing the database system under existing middleware. The committee believes that the SSA has underestimated the technological risks of this conversion approach and has not fully considered alternative approaches. The SSA's contractor involved in this effort and the SSA appear to have adopted the philosophy that any MADAM-to-DB2 conversion should have little or no impact on existing applications—that is, the applications would not be rewritten to take advantage of more contemporary software and hardware capabilities. This approach not only limits the functionality of those applications but compromises the design of the new database. The proposed approach would almost certainly cause poor performance, potentially lead to update anomalies, and almost inevitably be the source of numerous long-term complications. Moreover, studies made available to the committee indicate that the SSA's contractor did not fully consider large-scale relational database management systems (RDBMSs) other than DB2.

To be sure, replacing MADAM would be far from a simple undertaking. The committee believes that the SSA's current "halfway" strategy would likely result in a working system but that it would also introduce significant new challenges of its own. The committee believes that a strategy involving a total conversion of the databases (using a modern RDBMS) and a rewrite of the application software is likely to yield the best results in the long run. However, such a major rewrite of the application software is itself a risky undertaking. The SSA's predicament resembles that faced by a number of other large organizations and firms—such as banks and telephone companies—that have had to deal with large, critical systems that used obsolete technology. Like the SSA, they have had to modernize large custom systems that cannot easily be migrated using standard database-migration solutions. The experience of such institutions would be invaluable to the SSA as it seeks to navigate a similarly challenging course.

Without detailed information about the current nature and status of the SSA's MADAM conversion efforts, the committee did not reach conclusions as to whether any of the SSA's ongoing conversion efforts should stop. However, it would be a mistake for the SSA to rely on too narrow a range of technical expertise to plan and/or execute something as central to its operations as the MADAM conversion effort. Not relying on a broad range of technical expertise to architect such a massive conversion effort—and not to fully explore alternative approaches—will make the achievement of successful outcomes more difficult and less likely. In seeking broader technical advice as opposed to advice from a relatively narrow range of vendors and contractors, the SSA might, for example, establish an external advisory board—consisting of experts from the database software industry, several large commercial enterprises (banks and telecommunications companies) that have deployed large relational databases, and academics—to oversee the MADAM conversion effort. Given the national importance of the SSA and its computer systems, it should be possible to attract top technical talent to provide such advice.

Finding: In 1986, the Office of Technology Assessment alerted the SSA to the technology and technological risks inherent in its Master Data Access Method, or MADAM. Today, MADAM and the SSA's current data-management approach continue to pose increasing risks. The approach faces increasing limits in the availability of staff who understand and can support the SSA's technologically obsolete, custom solution. In addition, the approach precludes the use of valuable new technological capabilities and requires interruptions in service for batch updates, both of which impede the provision of desirable new e-services.

Recommendation: As it makes decisions about future directions for its database technology, the SSA should give considerable weight to the implications of those decisions for the effectiveness and efficiency of current and future electronic service delivery and should be open to the introduction of new technologies.

Finding: The scope and scale of the challenges that the SSA faces with regard to its database conversion strategy merit the input of a broad range of expertise.

Recommendation: In continuing to develop its conversion strategy and long-term services strategy, the SSA should draw on a broad range of technical expertise—including but not limited to database software experts, software engineers, software security experts, financial services experts, large-scale commercial service providers, and systems architecture experts—and put systematic mechanisms in place so that it can hear and learn from outside advisers.

TOWARD ORGANIZATIONAL TRANSFORMATION

The SSA has been slow to develop and implement strategic service-delivery plans, despite repeatedly being encouraged to do so by outside auditors and experts. The reasons for this are not entirely clear. One contributing factor may be a lack of sustained leadership commitment to this issue over the years. Another may be an organizational culture that is focused on personal, individualized customer service and has had a long tradition of understanding its public and how to serve it best—that is, through personal attention. The cultural mores within the SSA seem to equate electronic or online services with impersonal service without giving due consideration to the opportunity for the SSA to use electronic or online services to respond more quickly, to provide greater convenience, to enhance user satisfaction, to increase accuracy, and to reduce costs. A final factor may involve the prospect that organizational change would also require an updated employee skill mix to support the technological innovation required for the development and implementation of a service-delivery strategy that embraces online services.

Although charting a roadmap for the future under these circumstances may be daunting, the terrain is not completely unexplored and uncharted. The experiences of and lessons learned by other institutions should serve as a useful guide to the approaches that the SSA might take as it expands its electronic service offerings. Moreover, as the following subsections indicate, there are a number of areas in which changes to organizational practice and culture would greatly help the SSA develop and refine its service-delivery strategy.

Balancing Risks and Rewards

The committee's impression is that the SSA takes a conservative and cautious approach to service provision. Indeed, being subject to constant and intense scrutiny from the U.S. Congress and various federal oversight agencies not unexpectedly causes the agency to be reluctant to assume risk. Yet failing to make needed changes can also incur costs that can eventually outweigh the risks associated with making such changes.

Finding: The SSA may be missing important opportunities to make sustained improvements in its service delivery because of an overemphasis on the potential risks of modernizing its service-delivery strategy and a lack of emphasis on the long-term risks associated with *not* revamping that strategy.

Recommendation: When evaluating new electronic service-delivery initiatives, the SSA should when appropriate seek to balance risks and rewards by recognizing such upside benefits from automation as cost reduction, fraud prevention, and customer satisfaction.

Metrics

The SSA's own internal assessments of its strategies, initiatives, and deployments are ongoing. The identification of appropriate metrics and the continual gathering of adequate measurements needed to ensure effective decision making can be useful in such processes. When organizational goals can be systematically reduced to metrics that are quantifiable (although not necessarily directly monetized), such metrics and measures can be the key components in a program of continuous improvement, as they help demonstrate progress in achieving organizational goals such as meeting the various needs of diverse publics and user communities.

The committee understands that it seems easy enough to simply assert the importance of quantitative measures in managing toward meeting goals, but that putting in place the details of how to do so may be far more difficult. It seems most important, however, that the SSA adopt as a goal the movement toward basing its operations and decisions on this approach. The establishment of effective, agreed-on metrics and measures would provide immediate benefits for the better management of projects, and potentially they could better justify requests for funds for the deployment of electronic services.

In general, metrics and measures should stem from a careful examination of goals that are ultimately quantified. The units of this quantification would be used to identify the metrics. Implementing data-collection processes in the framework of those metrics would yield measures. Engaging in dialogue with managers from other organizations that manage in a metrics-oriented fashion and drawing from their "lessons learned" may be useful. In some cases, organizations find it helpful to engage the services of management consultants and/or to have ongoing dialogues with people who have implemented this approach in related institutions, such as large banks or brokerages.

Finding: The establishment of appropriate metrics and measures to evaluate the effectiveness of various services and delivery channels is an important component of an effective service-delivery plan.

Recommendation: The SSA should define and use metrics and measures to assess and improve its service delivery across all channels, including electronic services.

Partnerships

Large organizations such as the SSA are increasingly deciding that it is unsustainable to maintain in-house all of the competencies that they require in order to meet challenges that are growing in scope and number. Such organizations often make conscious decisions about which core competencies they will develop and nurture in-house and which they will outsource. Given the ongoing transition to and emerging demand from a variety of user communities for large-scale and highly effective electronic services, there may be increased opportunities that the SSA could explore in this area.

In the federal arena, there are precedents for partnering with the private sector. For example, the Internal Revenue Service (IRS) implemented the free file tax program with private-sector partners; in this case the IRS was able to develop the necessary partnerships within existing law and regulation, while in other instances legislative changes were sought and obtained. Other types of opportunities that might prove fruitful for the SSA include partnering with the states to enable the delivery of certified, electronic vital records evidence, promoting standards work involving data interchange, or increased cooperation with third parties regarding advice and counseling in the area of claims filing. Partnership opportunities for some of the SSA's activities may be possible within its current legislative and regulatory framework. In other cases, new authority may be needed in order to pursue partnership opportunities to enhance service delivery. In the area of electronic services, organizations have often found it useful to consider what their core competencies need to be and then to seek ways to partner effectively to fulfill other functions. In exploring potential partnerships, the committee suggests using an open process that engages beneficiaries, third parties, and other user communities as well as Congress to explore needs and alternatives.

Finding: There are opportunities for the SSA to partner with other agencies and third parties in ways that could provide mutual benefit.

Recommendation: The SSA should undertake to understand the identities, needs, and attitudes of its various user communities and should use that information to establish effective relationships and ongoing interactions with users, potential partners, and third parties. The SSA should explore partnering opportunities and identify the changes and initiatives that are necessary in order for it to enable appropriate interaction and cross-functionality with strategic partners and to support the exchange of data with other government agencies (both federal and state) while ensuring that appropriate security and privacy measures are in place.

CONCLUSION

The recommendations above are subordinate to what the committee sees as a more general and overarching need: that the SSA embrace change as a constant factor in the way that it does business. Particularly as change continues in the various domains of SSA involvement, it will become increasingly important for the SSA to devote attention and resources to being well informed about the nature and ramifications of these changes, both for itself and for its various user groups, including beneficiaries. Although the SSA does pay attention to change and seeks advice externally and internally, there are opportunities to use that advice more systematically as a basis for sustained and effective action. Broader and more systematic attention to anticipating and addressing change is warranted.

Finding: The SSA faces significant ongoing change—in terms of technology, demographics, and public expectations—as it carries out its activities, services, and interactions with a variety of user communities.

Recommendation: The SSA should embrace change as a constant. It should regularly evaluate emerging trends in such areas as technology (for example, database technologies) and business practices (for example, by learning from the experiences of financial institutions and moving toward the use of strategic partnerships for efficiency and effectiveness). It should also regularly evaluate the changing societal attitudes and expectations of its various user communities. The SSA should also institutionalize the formulation of strategies for addressing these trends.

Although the challenges outlined in this report are numerous and sizable, the committee is confident that they are not insurmountable for the SSA. Throughout the course of the committee's work, it has been clear that the SSA and its people are firmly dedicated to meeting beneficiaries' needs with enthusiasm and professionalism. Their dedication to their mission seems absolute and unwavering. This report is offered in the spirit of advice to dedicated professionals about how they address the opportunities that exist for meeting continually growing challenges.

1

Background and Current Context

This chapter provides some background on the Social Security Administration's (SSA's) mission and strategy and presents a brief overview of the rest of the report.

THE MISSION OF THE SOCIAL SECURITY ADMINISTRATION

The mission of the Social Security Administration is "to promote the economic security of the nation's people through compassionate and vigilant leadership in shaping and managing America's Social Security programs."[1] The agency's mission, along with its organizational structure and culture, has its roots in the Great Depression and President Franklin D. Roosevelt's broad initiatives for addressing economic insecurity and poverty.

Development and Expansion of the Social Security Act

In 1934, with the country traumatized by the Great Depression, President Roosevelt pressed forward to provide economic security for older Americans in the form of old-age benefits from an insurance system with near-universal coverage.[2] The original Social Security Act was signed by

[1]See Social Security Administration, "Information About the Social Security Administration," available at http://www.ssa.gov/aboutus/, accessed June 9, 2006.

[2]See, for example, Richard E. Neustadt and Ernest R. May, *Thinking in Time: The Use of History for Decision Makers*, New York: The Free Press, 1986, pp. 97-102.

the president on August 14, 1935. The 1935 act included two major provisions for the elderly:

- Title I (Grants to States for Old-Age Assistance) supported state welfare payments for the aged, and
- Title II (Social Security) provided benefits to workers upon retirement at age 65.[3]

The Social Security Act of 1935 provided *retirement benefits* for workers at age 65. Social Security benefits and programs were expanded through a series of legislative amendments to cover the following:

- In 1939: *dependent's benefits* for the spouse and minor children of a retired worker and *survivor's benefits* for the family of a covered worker who died before retirement,
- In 1950: increased benefit amounts and *cost-of-living allowance* (COLA) increases,
- In 1954 and 1956: *disability benefits*,
- In 1956 and 1961: *options providing for early retirement* at age 62 (with reduced benefits) for women in 1956 and for men in 1961,
- In 1965: *Medicare* (an SSA-administered health insurance program for people over 65), and
- In 1972: a modern *Supplemental Security Income* (SSI) program and yearly COLA increases.[4]

The SSA began as an independent agency at the subcabinet level. It became part of the new Department of Health, Education and Welfare in 1953; was made part of the new Department of Health and Human Services in 1980; and was returned to independent-agency status in 1995.[5]

The SSA no longer administers Medicare, which is now administered by the Centers for Medicare and Medicaid Services (CMS; formerly the Health Care Financing Administration), itself part of the Department of Health and Human Services. Although CMS is in charge of Medicare, the SSA provides substantial service-delivery support for the program. For example, applicants can receive general information about Medicare and its programs from the SSA and can apply for Medicare through the

---

[3]Adapted from Social Security Administration, "Social Security History" available at http://www.ssa.gov/history/briefhistory3.html, accessed June 9, 2006.

[4]Social Security Administration, "Social Security History," available at http://www.ssa.gov/history/briefhistory3.html, accessed June 9, 2006.

[5]Abridged from "SSA History," available at http://www.ssa.gov/history/orghist.html, accessed June 9, 2006.

SSA.[6] The SSA also provides substantial service-delivery support for the Medicaid, Railroad Retirement, and Food Stamp programs.[7]

Current SSA Programs

The Social Security Act, as amended, established the following three programs that the SSA currently administers:

- Old Age and Survivors Insurance (OASI),
- Disability Insurance (DI), and
- Supplemental Security Income (SSI).

The OASI program is financed by the OASI Trust Fund. According to the SSA, 92 percent of persons aged 65 or over in 2004—some 40 million people—were receiving OASI benefits as retirees, spouses, or other dependents; these benefits amounted to more than 50 percent of income for 65 percent of these beneficiaries.[8] In that same year, more than 158 million individuals earned benefits by paying Social Security payroll taxes, and the SSA paid more than $490 billion to more than 48 million people.[9]

The DI program is administered by the SSA. The SSA funds state-run Disability Determination Services (DDS). Although these state-run DDS agencies are federally funded and guided by SSA rules in their decision making, they hire their own staffs and retain a considerable degree of independence in how they manage their offices and conduct disability

[6]Medicaid is a different program, run by the states, that provides qualifying, low-income people with medical and hospital coverage. For more information, see http://www.ssa.gov/pubs/10043.html#part7, accessed June 9, 2006.

[7]Social Security Administration, *Strategic Plan: FY 2006-FY 2011*, January 2006, p. 2 (hereafter cited as Social Security Administration, *Strategic Plan: FY 2006-FY 2011*), available at http://www.socialsecurity.gov/strategicplan.html, accessed June 9, 2006.

[8]Social Security Administration, "Management's Discussion and Analysis," pp. 8-9 in *Performance and Accountability Report, FY 2005*, January 2006 (hereafter cited as Social Security Administration, *Performance and Accountability Report, FY 2005*), available at http://www.ssa.gov/finance/2005/MDA.pdf, accessed June 20, 2007. According to the SSA Web site at http://www.ssa.gov/deposit/DDFAQ898.htm (accessed June 20, 2007), "as of January 1999, 75 percent of all Social Security and SSI beneficiaries received their benefits by direct deposit."

[9]Social Security Administration, *Results at the Social Security Administration: Getting It Done*, August 2, 2005, p. 1, available at http://www.socialsecurity.gov/performance/results/results2005.pdf, accessed June 20, 2007. To qualify for retirement and survivor's insurance benefits, a worker born after 1929 must have paid Social Security taxes for at least 10 years (thus receiving 40 "credits") during his or her lifetime. Credits are based on earnings; in 2006, a worker could earn one credit for each $970 in earnings, up to a maximum of four credits a year. Benefits are payable to workers upon retirement at age 62 or later; benefits are also paid to certain dependents and survivors.

determinations.[10] To qualify for the DI program, an individual must be eligible through recent covered work before the onset of disability. The number of recent credits required for eligibility increases with age. "Disability" requires that a worker have a mental or physical impairment that has lasted (or is expected to last) more than 12 months or that is expected to result in death (the DDS agencies perform the medical review function and are responsible for making the medical determinations under SSA's direction). DI benefits can be paid to disabled workers and eligible family members as long as the individual is disabled and does not perform "substantial gainful work." The SSA periodically reviews the disability status of beneficiaries. The SSA also offers programs to provide incentives for individuals who want to try to return to work. According to the SSA, in 2005 SSA DI benefits replaced about 44 percent of the income of a "medium income" disabled worker. DI benefits were paid to about 8 million beneficiaries.[11]

SSI is a needs-based program financed from general tax revenues, in contrast to the retirement and survivor's insurance benefits provided under the OASI program and the disability insurance benefits provided under the DI program. It is designed to provide benefits to aged adults and to blind or disabled adults and children with limited income and resources. For adults, the SSI definition of "disability" and the SSI disability review procedures are the same as those for the DI program, except that different rules apply for statutory blindness. Different definitions apply for children. There are provisions and incentives intended to encourage people receiving SSI benefits to work. Because it is a needs-based program, SSI has ongoing requirements for recipients to submit and the SSA to process monthly information about income and resources. This information impacts benefit amounts and continuing eligibility. The full SSI benefit is designed to be equivalent to about 73 percent of the federal poverty level for an individual and about 81 percent for a couple. In 2005, the federal poverty level was defined as $9,750 for an individual and $12,830 for a couple. Most states supplement the federal SSI benefit. According to the SSA, in 2005, 4.6 million individuals received SSI benefits only; 2.5 million received concurrent SSI and OASI/DI benefits.[12]

---

[10]U.S. General Accounting Office, Testimony Before the Subcommittees on Human Resources and Social Security, Committee on Ways and Means, U.S. House of Representatives, *SSA Customer Service: Broad Service Delivery Plan Needed to Address Future Challenges*, February 11, 2000, available at http://www.gao.gov/archive/2000/h100075t.pdf, accessed June 20, 2007.

[11]Social Security Administration, *Performance and Accountability Report, FY 2005*, pp. 8, 10.

[12]Social Security Administration, *Performance and Accountability Report, FY 2005*, pp. 2, 11.

Although the most visible users in these programs are members of the general public, users in addition to individual beneficiaries also have interests in the SSA's varied programs. Most notably, state governments, as joint administrators of programs such as DI, have important stakes at least in how various databases of information are maintained and made accessible: proper access directly affects the timeliness and accuracy of processing disability claims and providing benefits to those who are found eligible. See Chapter 4 for more on external partnering.

SSA Business Products and Processes

According to the SSA, the OASI, DI, and SSI programs "touch" over 95 percent of the American public at various points during their lifetimes. The reasons include the following: filing for a Social Security number (SSN), establishing a record of earnings, retiring, becoming disabled, suffering the loss of a spouse or parent, and/or being unable to meet basic financial needs as an older American. The SSA views these as the critical points at which members of the public interact with the agency by seeking information, applying for benefits, or reporting "post-entitlement" changes. See Table 1.1 for a sense of the scale of the SSA's activities across the three programs.

TABLE 1.1 Social Security Administration Activities in Fiscal Year 2005 Across Three Programs: OASI, DI, and SSI

Activity in Fiscal Year 2005	Scale
Benefits paid	Almost 53 million people per month
Eligibility determinations made	8 million new claims
Decisions made	1.6 million hearings and appeals
Continuing disability benefits reviewed	1.5 million reviews
New and replacement Social Security cards issued	17 million cards
Worker's earnings records processed	257 million items
Calls received at the SSA 800-number	56 million calls
Social Security Statements issued	142 million statements

NOTE: OASI, Old Age and Survivors Insurance; DI, Disability Insurance; SSI, Supplemental Security Income.

SOURCE: Social Security Administration, *Performance and Accountability Report, FY 2005*, January 2006, p. 12, available at http://www.socialsecurity.gov/strategicplan.html, accessed June 9, 2006.

As it carries out its mission through these business processes, the SSA emphasizes "Service, Stewardship, Solvency, and Staff" as shorthand for its commitment to do the following:

- Give the American people the excellent service they expect and deserve,
- Ensure the highest level of program integrity through sound fiscal stewardship,
- Ensure the program's financial solvency for future generations, and
- Maintain high-quality staff committed to organizational excellence.[13]

These are the top-level, strategic commitments that are reflected in the agency's formulation of strategic goals and objectives and performance indicators. In fiscal year (FY) 2005, according to the SSA, 78 percent of its operating expenses, some $7.4 billion, were used in support of the service goal.[14]

The SSA has defined five core business processes to facilitate the planning and managing of the delivery of services to beneficiaries:

- Issuing SSNs (enumeration),
- Establishing and maintaining individual records of earnings,
- Processing benefits claims,
- Maintaining post-entitlement records of changes and reviews, and
- Informing the public.[15]

These core business processes cross program and organizational lines within the agency.[16] For example, processing benefits claims applies to all programs for which the SSA is responsible. Although there are variations among the claims processes for the respective programs, the essential claims process is the same. Similarly, the basic post-entitlement business process applies to all programs, again with some variations among them. The same basic communications strategies, tools, methods, and so on are used to inform the public about each of the programs, although targeted audiences, emphases, and media might be used to suit the program need. In all cases, executing these processes also entails crossing organizational lines.

Although the focus of each of these processes is to serve members of the public, each also entails important interaction, either current or

[13]Social Security Administration, *Strategic Plan: FY 2006-FY 2011*, p. 1.
[14]Social Security Administration, *Performance and Accountability Report, FY 2005*, p. 15.
[15]Social Security Administration, *Strategic Plan: FY 2006-FY 2011*, p. 2.
[16]Social Security Administration, *Strategic Plan: FY 2006-FY 2011*, p. 2.

potential, with other communities. Thus, for example, while the issuance of SSNs is a service to individual members of the public, the verification of SSNs is an important service to business and other governmental agencies. Serving these additional users entails responding to many millions of additional service requests.

INTERNAL ORGANIZATION OF THE SOCIAL SECURITY ADMINISTRATION

The Social Security Administration has a staff of more than 65,000 employees spread throughout a network of some 1,500 offices. In addition to the Baltimore, Maryland, and Washington, D.C., headquarters, these offices include the SSA's 10 regional offices; 7 processing centers; more than 1,300 field offices (which, among other things, handle application intake); 36 teleservice (800-number) centers; and hearings offices. In addition, the SSA provides policy, administrative direction, and funding for the 54 state-run DDS agencies, which have more than 16,000 employees.[17]

Public contact with the SSA is primarily conducted face to face at the field offices, by phone through the teleservice centers, or by mail (for example, Social Security Statements are automatically mailed yearly to workers over age 25). Over 60 percent of the SSA's employees deliver direct service to the public, mainly in field offices and teleservice centers; another 30 percent in the regional offices, processing centers, and headquarters, providing direct support to those front-line workers.[18] However, the SSA is also seeking to use online information and online interactions with the public, both to obtain the efficiencies that technology can offer and to achieve the increases in service capacity that will be required in order to handle the growing baby boom-related workload.[19] The SSA also looks to online interactions as a way to help in providing services to its wider user communities.

The SSA is headed by the Commissioner of Social Security.[20] In addition to the Office of the Commissioner, there are 15 major headquarters

[17]Social Security Administration, *Performance and Accountability Report, FY 2005*, pp. 13-14.

[18]Social Security Administration, *Performance and Accountability Report, FY 2005*.

[19]The SSA is facing a baby boom "retirement boom" increase in workload just as its own workforce is becoming eligible for retirement: more than 40 percent of the SSA's employees are expected to retire by 2014. The baby boom generation has already entered the "disability-prone" years; as a result, the SSA expects its DI rolls to increase by 35 percent between 2002 and 2012 (Social Security Administration, *Strategic Plan: FY 2006-FY 2011*, pp. 8, 41).

[20]The current Commissioner is Michael J. Astrue, whose term began in February 2007. When this study was initiated, the Commissioner was Jo Anne B. Barnhart, whose term began in 2001. The first Commissioner was Arthur J. Altmeyer, who served from 1946 to 1953. See http://www.ssa.gov/history/commissioners.html, accessed June 9, 2006.

components, organized along functional lines. Each component is headed by a Deputy Commissioner or other senior-level official. (See Appendix C for brief descriptions of the functions of these major offices.)

SSA Organization and E-Government Services

Just as the SSA's defined business processes cross lines of benefits programs, the SSA's management structure crosses lines of both program and delivery channels. This subsection is based on material and quotations from the "Organizational Structure of the Social Security Administration" at the SSA's Web site (http://www.ssa.gov/org/ssaorg.htm). Quotations below are from the SSA's organizational element descriptions. With respect to the SSA's e-government initiatives, the focus of this report, responsibilities for software, hardware, and support for providing electronic information and services are split across several Deputy Commissioners' domains:

- The Office of Automation Support in the *Office of the Deputy Commissioner, Operations* (ODCO) is responsible for "integrating service delivery and employee concerns with modern technology." This office determines Operations' requirements for software, hardware, and electronic service-delivery support. Also in Operations, the Office of Electronic Services (OES) is the lead for "development and implementation of electronic services."
- Under the *Chief Information Officer's* leadership, OES also works with other federal agencies on interagency electronic service-delivery initiatives.[21]
- The *Office of the Deputy Commissioner, Systems* (ODCS) contains the Office of Telecommunications and System Operations, which is responsible for the management, operation, and maintenance of the computer systems and networks on which both e-government and voice applications run. It also contains the Office of Systems Electronic Services, which directs the development of the software that supports electronic service-delivery initiatives; the Office of Disability Systems, which develops, implements, and maintains electronic systems to support disability programs (such as the "eDIB" initiative[22]); and the Office of Enterprise

[21]See Office of Management and Budget, *E-Government Strategy: Simplified Delivery of Services to Citizens*, February 2002, available at http://www.whitehouse.gov/omb/inforeg/egovstrategy.pdf, accessed June 20, 2007.

[22]The eDIB system replaces the paper SSA disability folders with electronic records accessible to all case-processing personnel officers across the country. The electronic folder addresses the problem of lost paper folders. Moreover, it eliminates delays in transferring paper files and can be used to support electronic hearings during the appeals process. See

Support, Architecture and Engineering (OESAE), which "identifies the strategic information technology resources needed to support SSA business processes and operations and the transition processes for researching, demonstrating and implementing new technologies in response to the Agency's strategic vision." OESAE includes the Division of Data Base Systems—among other things this division develops and maintains the crucial Master Data Access Method (MADAM) software that manages the benefit programs' master files.

- The *Office of the Deputy Commissioner, Communications* (ODCComm) is responsible for the creation, development, evaluation, and oversight of all internal and external SSA communications, as well as its public affairs and public information activities. Within the Office of Communications, the Office of Communications Planning and Technology is the focal point for the development, clearance, and placement of content material on the SSA's official Internet/Intranet Web sites and is responsible for the development, content, and coordination of the SSA's internal and external Web marketing activities. The Office of Communications has responsibility for setting policy, for determining what information content is posted on the SSA Web site, and for maintaining the "look and feel" of the entire site.
- Within the *Office of the Deputy Commissioner, Disability and Income Security Programs* (ODCDISP), the Office of the Associate Commissioner for Disability Programs is responsible for the development, coordination, and oversight of disability policies, procedures, and process requirements supporting the creation of a paperless disability claims process.

These split responsibilities for electronic information and services and their implications are discussed further in Chapter 4.

The SSA, like other government entities, seeks ways to improve the delivery of information and services while also reducing costs.[23] To accomplish these effectiveness and efficiency improvements, the SSA and others are turning to more expansive applications of information technology (IT) and looking to electronic services as a way to help in providing its services to its users. (See Appendix E for a short history of federal e-government activities to provide a broader federal context for the SSA's activities in this area. Also see Box 1.1 for a brief discussion of terminology.)

"Agency Challenges," pp. 25-26 in Social Security Administration, *Performance and Accountability Report, FY 2005*, available at http://www.ssa.gov/finance/2005/Agency_Challenges.pdf, accessed June 12, 2007.

[23]One key benefit of providing electronic access to SSA transactional services is likely to be a substantial reduction in error rates and, subsequently, in associated costs. For example, presumably an SSA user or beneficiary is less likely to enter his or her own name and address incorrectly than is a call-center person who only hears it or a key entry person who is transcribing it from a handwritten entry on a paper form.

BOX 1.1

"Electronic Services" Meant to Be Construed Broadly

Throughout this report, several terms are used that are often perceived as nearly interchangeable: "online services," "e-services," "e-government," "electronic services," and so on. Although there are subtle distinctions among all of these, the committee's preferred term is "electronic services"; it is meant to encompass all of the above. That term was chosen in part because it is the most generic and all-encompassing term, and it also closely mirrors the terminology used currently by the Social Security Administration. It is meant to encompass a broad vision of service delivery—not just to include interactions that take place solely through the use of the Internet, although that is a primary focus, but encompassing others as well.

THE AGENCY'S BROAD BASE OF USERS

The SSA's clients and users are not just individual beneficiaries but also include federal government agencies and state governments (for example, the U.S. Internal Revenue Service [IRS], U.S. Department of Labor, and state offices of vital records—see below); third parties (for example, payroll services); representatives assisting beneficiaries (for example, attorneys or representative payees); internal agency users of electronic services (for example, field office workers); and external users of electronic services (for example, community service agencies and social science researchers). Consideration of how the SSA should position itself to provide electronic services in the future should take into account the needs of these various communities and whether they are to be addressed individually or comprehensively. Owing to limited resources and scope, this study did not examine the full range of needs of all of these users. The following subsections indicate the character of some of these user needs.

Federal Agencies and State and Local Governments

Increasingly, in both e-commerce and e-government, organizations such as the SSA partner with a variety of other organizations to deliver electronic products and services. The SSA already does this to some extent through its participation in some of the federal e-government initiatives that require interagency and intergovernmental coordination and cooperation. For instance, the eVital project involved the SSA's working with the organization that represents state vital statistics agencies—the National Association for Public Health Statistics and Information Systems—to streamline the process of reporting and verifying state-level birth and death data electronically.[24] In this instance, the SSA found a community that shared an interest in speeding up and increasing the accuracy of data needed by both partners to fulfill their missions. More specifically, automating the data exchange between state and federal agencies reduced costs, minimized errors from manual processes, and reduced unnecessary and sometimes fraudulent benefits payments.[25]

The activity just described also points out that a government-to-government e-government project can benefit both the government and SSA beneficiaries without the SSA's necessarily having to interact directly with beneficiaries. There are challenges of course. As an obvious condition of such interagency and intergovernmental data exchanges, the SSA will want to ensure that the organizations with which it intends to exchange data are willing to comply with the privacy and security requirements that surround data maintained by the SSA on behalf of workers, beneficiaries, and other users. Toward this end, the SSA likely will engage in discussions with potential data-sharing partners and work with them to reduce risks, thereby obtaining important new benefits.

There are risks in failing to take advantage of important new technologies, interactions, and opportunities. To be sure, any data-sharing or data-exchange agreement would have to be cost-effective enough for the parties to agree to it. In addition, many states and most counties may not have automated records or only have automated records from fairly recent times. Resource constraints may hamper the ability of some states and counties to create electronic databases despite the obvious long-term benefits to be gained over time from the up-front investment well beyond merely exchanging the data with the SSA.

As noted previously, the SSA is required to share information with a variety of other federal government agencies. Considerable exchange of information is required with the IRS in particular, but exchanges with the Department of Veterans Affairs, for example, are also an important part of the SSA's role and mission. There may be increased interactions in the future with the Department of Homeland Security and other agencies as well. Because of this, the SSA will need to position itself to ensure that these intra-federal-government transactions are as prompt, efficient, and yet secure as possible. An ongoing consideration will be how to maintain appropriate privacy protection while meeting emerging demands. The creation of transaction modes that are convenient for all agencies will

---

[24]For more information, see http://www.ssa.gov/pressoffice/evital-pr.htm, accessed June 12, 2007.

[25]For more information, see http://www.whitehouse.gov/omb/egov/c-2-4-evital.html, accessed June 12, 2007.

clearly be important. But it is also important that the SSA and its federal partners not overlook the need to ensure that underlying data repositories are designed and implemented so as to facilitate and support sufficient privacy protection, access, security, and sharing modes that serve all participants as effectively as possible.

The SSA interacts with state governments in a number of ways, most notably in the handling of Disability Insurance cases. Because of the complexity of disability claims cases and the number, size, and diversity of the data files that comprise them, the demands already being placed on the SSA by the state governments are considerable, and they must be expected to grow in the coming decades. The nature of the data that the SSA must share with the states can be quite sensitive, placing considerable demands for protection of privacy on these transactions. The SSA has developed a system (the eDIB, or Electronic Disability, system) that is intended to improve the quality of its interactions with state-run DDS agencies in what appears to be recognition of the importance of this community in the SSA's future e-government plans.

Third Parties

Third-party entities can be potential service-delivery partners, and there are likely opportunities for the SSA to seek initiatives that provide mutual benefits to both organizations and their shared user bases. One challenge is that there are no clear-cut rules for initiating or managing such partnerships for federal agencies.[26] This lack of clear-cut rules also means, however, that agencies have wide discretion to create partnerships with third parties for electronic product and service delivery. There are models of innovative e-government partnership, with the IRS's Free File being the most visible example,[27] that can serve as both inspiration and a source of lessons learned. Mindful of privacy and security considerations—which the committee acknowledges can be considerable—the SSA would be well served to explore a broad set of possible partnerships to reach the widest set of users and beneficiaries.

Various types of third parties already interact with the SSA. Payroll services companies, for example, are quite prevalent in the U.S. economy, providing to private companies of all sizes support for the preparation of their payrolls. Included in this support are the collection of Social Security

---

[26]S.H. Holden and P.D. Fletcher, "The Virtual Value Chain and E-Government Partnership: Non-Monetary Agreements in the IRS E-File Program," *International Journal of Public Administration* 28(7-8):643-664, 2005.

[27]Information is available at http://www.irs.gov/efile/article/0,,id=118986,00.html, accessed June 12, 2007.

deductions and the preparation of the required Social Security reports. Attorneys specializing in the preparation of disability claims are also an increasingly noticeable presence in our society, and they have substantial interactions with the SSA. Representative payees—friends, family, or other parties who help beneficiaries manage their Social Security payments if the beneficiaries are not able to do so themselves—are another example of third parties that have substantial interactions with the SSA. Many of these third parties already help potential and current beneficiaries with SSA interactions and, as a result, already require the SSA's attention and resources. The role of third parties intermediating interaction between the SSA and its constituents is already present and likely to increase—at least in quantity if not in kind. The SSA will need to decide how to support third parties in the service-delivery process—while managing the SSA's appropriately stringent privacy and security requirements[28]—and what the potential of these individuals and groups is for assuming an important role in service delivery working with the agency.

THE SOCIAL SECURITY ADMINISTRATION'S E-GOVERNMENT STRATEGY DOCUMENT

As part of the study process, the Committee on the Social Security Administration's E-Government Strategy and Planning for the Future was asked to react to the SSA's "E-Government Strategy document."[29] Early in the study process, briefings from the SSA and others and an examination of documents provided by the SSA led the committee to conclude that the SSA faces fundamental challenges that, unless properly addressed, would significantly hinder *any* strategy for implementing electronic services. In the committee's view, electronic services are best examined in the context of the SSA's overall service-delivery strategy. Therefore, in the view of the committee, given available resources, an extensive focus on this one particular document would not have been the best use of committee efforts or best serve the SSA or the public. In essence, this entire report is the committee's response to the early strategy documents and the subsequent input and briefings that it received. Below is a brief discussion of the strategy document itself.

---

[28]Recent losses of personal data by the Department of Veterans Affairs, credit reporting agencies, and others highlight the importance of privacy and security considerations.

[29]This document, Social Security Administration, "E-Government Strategy: Meeting Expectations in a Changing World," dated July 22, 2004, was provided to the committee at the start of the study process.

The Strategy Document

The SSA describes the motivation for the "E-Government Strategy" document as follows:

> To guide the future development of electronic services and plan to meet the performance measure [targets set as part of the SSA's FY 2005 Agency Performance Plan development process], agency executives have established the SSA E-Government vision and goals along with the specific projects to achieve the goals within 3 to 5 years.[30]

The SSA sets its context for action in terms of the President's Management Agenda (PMA)[31] focus on "citizen-centered" use of information technologies to provide "high quality service, cost reduction, improved access to services and government accountability."[32] The SSA also notes the profound effect that the forthcoming baby boom retirement wave will have on its workload, as well as changing public expectations and behaviors with respect to online information and services. The e-government vision as outlined in the strategy document is to provide easy-to-use, secure, and cost-effective e-government services to individuals, businesses, and other government agencies by 2009, so that these clients and other users can conduct most of their business with the SSA electronically.

The PMA separates e-government projects into four cross-agency project portfolios: government to citizen (g2c), government to business (g2b), government to government (g2g), and internal efficiency and effectiveness (iee). The strategy document describes the SSA's progress in implementing discrete projects in each segment, including the Social Security Online Web site and the Internet Social Security benefit application (g2c), electronic wage reporting and the SSN verification service (g2b), data exchanges for secondary payer and veterans' benefits matching (g2g), and travel bookings (iee). It then highlights selected projects, such as eDIB, supporting the agency's e-government goals and lists e-government goals linked to three of the agency's four strategic goals (Service, Stewardship, and Staff). For example:

---

[30]Social Security Administration, "E-Government Strategy: Meeting Expectations in a Changing World," July 22, 2004, p. 4.

[31]For more information on the PMA generally, see http://www.whitehouse.gov/omb/budintegration/pma_index.html, accessed June 12, 2007. The SSA leads the eVital project (in the government-to-government portfolio described above); the goal is to establish common processes for federal and state agencies to collect, process, analyze, and share birth- and death-record information.

[32]All quotations in this subsection are taken directly from the "E-Government Strategy" document (see footnote 29, above).

- "Better identify SSA interactions that are appropriate for self-service" is an element in support of the Service goal (the lead is ODCO);
- "Develop an integrated policy framework on privacy, security, and disclosure," "Establish new g2b partnerships with our consultants and with interest groups," and "Manage and measure the cost-effectiveness of a portfolio of e-Gov products and services as opposed to an isolated focus on single Online products/services" are elements in support of the Stewardship goal (the leads are the Office of the Chief Information Officer [OCIO]/ODCS and ODCO; ODCO and ODCComm; and ODCO, respectively); and
- "Structure programs and policy to better fit the electronic world while continuing to maintain the integrity of the programs" is an element of the Staff goal (the lead is the E-Government Executive Council).[33]

The strategy document describes the governance and organization of the SSA's e-government activities as spread across the E-Government Executive Council (to provide leadership at the Deputy Commissioner level), the Associate Commissioner E-Government Steering Committee (to develop and monitor the implementation of the strategy), and the relatively new position of the Chief Information Officer (responsible for the e-government portion of the PMA and for ensuring that information technology and information resources are acquired and managed according to the Information Technology Management Reform Act of 1996, passed as part of the National Defense Authorization Act for 1996 [Public Law 104-106] and the E-Government Act of 2002 [Public Law 107-347]).

Brief Assessment

Over the course of the study, the committee examined the SSA's history, current status, and strategic plans for providing electronic information and services to its various user groups (see Box 1.2 for some examples of these users and uses). The resulting report is not so much a reaction to what is in the SSA's strategy document as a reaction to what is not included. In isolation, and at first glance, the agency's e-government goals—to employ a citizen-centered approach, to ensure privacy and security, to pursue partnerships, to achieve cost-effectiveness, to align the organization for progress—seem reasonable. However, in the committee's view, this approach will not take the agency where it needs to go. The strategic goals and projects do not break out of the status quo organizational culture and the highly cautious approach to adopting and

---

[33]Social Security Administration, "E-Government Strategy: Meeting Expectations in a Changing World," July 22, 2004, pp. 11-12.

deploying contemporary IT. They do not set the SSA on the path that it needs to be on in an environment of continuous technological and societal change. An implicit theme in this report is that the agency's technological underpinnings and its organizational culture should each be examined in order to ensure that all are poised to support the shift toward a culture of continuously striving to meet effectively the manifold pressures for change. The committee's vision is of an SSA that is proactive and thriving in an environment of continuous technological and societal change. The committee believes that this report can be useful in helping the agency's dedicated employees reach that vision.

OUTLINE OF THE REPORT

Beyond the Summary (which includes all of the findings and recommendations found in the body of the report) and the background and current context provided in this chapter, the rest of this report examines and assesses the SSA's medium- and long-term strategy for electronic services, including technological assumptions, operational capabilities, functional requirements, and future goals. Chapter 2 describes current electronic services offered by world-class financial institutions. In the committee's view, the experiences of financial institutions are a source of important lessons and insights for the SSA (and other agencies) seeking to meet modern expectations and requirements. Chapter 3 provides a brief overview and assessment of the SSA's current technological infrastructure and organizational approach and summarizes relevant technological and demographic trends, along with their implications for the effective provision of electronic services. Chapter 4 provides an assessment of the SSA's current organizational culture and structure and addresses the impact of culture and structure on the SSA's IT choices and service delivery. Chapter 4 concludes with a discussion of opportunities for change. The report outlines a variety of ways in which the SSA can position itself to meet the demands of the future and effectively integrate electronic services into its long-term service-delivery strategy.

BOX 1.2
Electronic Services at the Social Security Administration: Users and Uses

Throughout this study, the Committee on the Social Security Administration's E-Government Strategy and Planning for the Future has been mindful of the broad range of current and potential users of Social Security Administration (SSA) electronic services, as well as of the variety of uses that online capabilities do and can enable. The following examples present ways in which an increasingly effective www.ssa.gov system can help serve the spectrum of SSA client and user communities. It is not intended to be exhaustive, but rather to illustrate capabilities that are available now[1] or that could be made available in the future. By presenting these examples, the committee does not intend to suggest or imply that the SSA should provide only electronic services in any particular program or to a certain group—indeed, multiple channels will always need to be maintained, as there will always be some population that is unable or unwilling to use electronic services.

• *A worker approaching retirement age* can now use the SSA's Retirement Planner tools, available at http://www.ssa.gov/retire2/, to find his or her retirement age, to see the trade-offs between retirement age and benefit amount, to learn how work after retirement and other benefits affect SSA payments, and to complete most of the paperwork needed to apply for retirement benefits. At present, applicants must mail originals or certified copies of supporting documents such as a birth certificate to the SSA or take them to an SSA office, even if the application is made electronically. In the future, if applicants were able to include identifying information about their birth certificates (or circumstances of birth) in the electronic application and if this information was sufficient for use by the SSA to obtain certified birth information through a partnership with the states, then the application process could be entirely electronic, faster, and would not generate visits to local SSA offices.

• *A retiree* can find information about the Medicare Prescription Drug Plan now at http://ssa.gov/prescriptionhelp/.

• *The father of a low-birth-weight child* can go to http://www.ssa.gov/d&s1.htm now for information on Supplemental Security Income (SSI) eligibility and on how to apply for SSI disability benefits on behalf of his daughter.

• *A newly disabled worker* can find information now about the SSA programs for which he or she may be eligible, detailed information about the Disability Insurance (DI) and SSI programs, fill out an online disability application, and get an Adult Disability Starter Kit to help prepare for the disability interview. Medical providers can submit supporting information online through the SSA's Electronic Records Express Web site, or fax the records to the SSA or the state Disability Determination Services (DDS) agency. Information about Electronic Records Express is available now at http://www.ssa.gov/ere/index.html.

• *A self-employed person* can use his or her Social Security number (SSN) as a taxpayer identification number, can report annual income and expenses by filing Schedule C, and can calculate the amount of self-employment tax owed by filing

[1]Online resources indicated as "now" available were available at the stated locations as of April 17, 2007.

Schedule SE, both of which are attached to his or her current federal income tax Form 1040. The self-employment tax includes both Social Security and Medicare payments. The IRS now sends a record of this Social Security tax payment to the SSA so that it can be added to the individual's earnings history.

• *Service-provider companies* offering accounting and payroll services can help their clients to comply with requirements for SSN verification and electronic wage and W-2 reporting. Employers and their authorized representatives can use the suite of services available now at the SSA's Business Services Online site, http://www.ssa.gov/bso/bsowelcome.htm.

• *Attorneys, advocates, and other third-party representatives* can assist clients who have a variety of disabilities that might entitle them to SSA disability and income support benefits. Their primary services are to help beneficiaries negotiate the complex interactions between the SSA's program rules and those of other federal agencies and state and local social services agencies. In most cases, clients have signed a power of attorney to recognized third-party representatives to represent them before the SSA to help ensure that they receive the full benefits to which they are entitled. Information for advocates, attorneys, and third-party representatives is now available at http://www.ssa.gov/onlineservices/thirdparty.htm and http://www.ssa.gov/thirdparties.htm.

• In the future, a *Social Security claims representative* (CR) in a local field office could use a suite of electronic services to develop the required electronic evidence to support disability or retirement benefits claims as well as continuation of disability benefits. The CR typically obtains prior-year earnings and tax information from the Internal Revenue Service. The CR also needs to get information from a variety of state databases to ascertain proof of age, marriage, and relationship to pre-teen children in the family. Determining eligibility might also require information not currently readily available to the SSA. Automated tools to facilitate this process would help ensure that beneficiaries are more likely to receive the benefits to which they are entitled more quickly and with fewer chances of error. Currently, only parts of this process are done electronically.

• *A woman planning her wedding* uses a wedding-related Web site with various "to-do" lists. She sees that, because she plans to change her name after she is married, she will need a new Social Security card in her married name. She uses a link on the site to go to www.ssa.gov and reads why it is important to get the new card now, finds a link to a form to fill out, and sees instructions on how to send or take the supporting documents to her nearest Social Security office. She downloads the "Application for a Social Security Card" (Form SS-5) from the SSA Web site available now at http://www.socialsecurity.gov/online/ss-5.html and begins to fill it out. Reading the instructions, she sees that she needs to show proof of citizenship, proof of identity, and proof of name change to the SSA, along with the completed form. She does not want to mail those documents to the SSA, so she uses the SSA Web site Local Office Search available now at https://s044a90.ssa.gov/apps6z/FOLO/fo001.jsp to locate the closest office. She realizes that she will have to leave work early sometime after her honeymoon so that she can take her marriage document, driver's license, and Form SS-5 to the SSA office. In the future, if she could securely send certified, electronic copies of her SS-5 form to the SSA, along with sufficient identifying information pointing to her certified identification and name change documents, she could avoid making a trip to the local SSA office.

2

Lessons from Electronic Services in Financial Institutions

Although much about the Social Security Administration (SSA) is distinctive, in some ways the SSA, its activities, its information technology (IT) needs, and its operational characteristics are analogous to other organizations outside the federal government. This chapter explores that premise and seeks to identify ways in which these organizations and institutions outside the federal government are dealing with some of the challenges that currently face the SSA in the realm of electronic services.

The committee recognizes that the SSA might not be able to—or want to—use "as is" all of the approaches and solutions pursued by other institutions. Some activities that citizens conduct with the SSA involve one-time, very personal, potentially traumatic events in their lives—at such times people desire the personal contact and support that traditional service-delivery options offer. Another stark distinguisher is the SSA's funding model. The agency has a budget limited by the congressional appropriations process. It is required to balance its spending of the appropriated funds. Given the federal context in which it operates, it cannot, as a general rule, raise new funds by attracting more business, nor can it easily justify exploring, then abandoning new services on the basis of a trial offering of new services as the financial industry might do. In practice, new services must readily be demonstrable to be in the public interest, and the benefits of providing those services should outweigh the benefits of developing other services competing for the same limited pot of funds.

There is nonetheless substantial value in considering what might be learned and what might be adapted or applied as a result of studying organizations that only provide somewhat-analogous services—if more-routine and less-personal interactions can be accomplished efficiently and with less staff overhead through the use of electronic services, that could, as an additional benefit, help to make more resources available for when more-personal service is needed. This chapter examines large-scale financial institutions having high-volume interactions with large numbers of individuals. The financial services community has been one of the most aggressive and competitive in using IT and electronic commerce. While the SSA is, appropriately, neither as aggressive nor competitive, there are ways in which financial institutions are dealing with situations that are relevant for the SSA. In addition, unlike more recent Internet-based companies such as Amazon or eBay, major financial institutions have had to undergo a transformation away from primarily bricks-and-mortar-based organizations to take advantage of and move into electronic service provision, a transformation much like what the SSA faces. At the same time, there are few "online-only" banks, meaning that most financial institutions have had to expand the kinds of channels through which they offer services, not replace them. Learning from their experience of that transformation may also be instructive for the SSA.

THE TRANSFORMATION IN FINANCIAL INSTITUTIONS

Just two decades ago, banks' interactions with their retail (individual) customers were almost exclusively walk-in or telephone transactions. Deposits and withdrawals were generally carried out in person, and account statements were printed on paper and sent through the mail, usually monthly. All of that started to change with the widespread deployment of automated teller machines (ATMs) and the creation of centralized call centers. In the 1980s, many services were introduced through proprietary services delivered to personal computers, screen phones, and television sets. However, the story of mass adoption, as well as economically feasible delivery of these services, begins in the mid- to late-1990s with Web-based delivery, fueled by growing public use of the Internet. The net effect of these changes has been to alter the entire character of the retail banking industry dramatically and to transform the way in which it is both used and perceived by its customers. Although not every bank has been as aggressive in all dimensions, there are lessons to be learned from best practices in the aggregate in the industry.

Some aspects of the role, activities, and operations of the SSA have much in common with those of a large commercial bank or brokerage house that maintains accounts and receives and makes payments. These institutions usually conduct transactions in multiple customer segments (individual retail customers, businesses, other financial institutions, and so on) and through multiple service channels (online, through ATMs and branch offices, through call centers, and so on). For reasons of cost and competitiveness, most such institutions have seen increasing value in emphasizing online customer channels to their services. From a coarse technical feasibility perspective, the databases that are maintained by private-sector institutions are now quite comparable, both in size and in transaction volume, to the databases for which the SSA is responsible (see Chapter 3). At least one major credit card issuer has 170 million open accounts, comparable with the SSA's approximately 140 million Social Security statements issued annually. These observations suggest that the experience, market research, and product-refinement knowledge accrued by these financial institutions during the past 15 years, as well as the set of practices and approaches to effective information- and service-delivery capabilities and customer service that they have developed, can be strongly relevant to the SSA. An examination of the commercial financial services industry offers relevant insights and lessons learned for effective electronic services. These can be useful to the SSA and other government agencies seeking to make more comprehensive and effective transitions to online services.

Online banking, in particular, seems to be a relevant, if not completely analogous, success story. A segment of the public has embraced the convenience that comes from immediate access to virtually up-to-the-minute information about personal finances 24 hours a day, 7 days a week (24/7).[1] Bills can be paid online and money can be transferred at all hours. Banking customers now expect that they can track the flow of money both into and out of their accounts at all times. Thus, 24/7 access has become the expected norm. Given that the SSA offers some similar types of services—albeit usually at a different frequency (monthly account changes instead of daily or hourly changes, for example)—this expectation is a reality that the SSA must confront. The SSA's current clientele is generally older, less abled physically and/or cognitively, and less financially well off than the general population of online bill payers. However, as the current population of online bill payers ages and starts using SSA services, their expectations will likely transfer to the SSA. Unlike today's population of people

[1]According to the Pew Internet and American Life survey of online banking in 2005, "fifty-three million people, or 44% of Internet users and one-quarter of all adults, now say they use online banking. Those figures amount to an increase of 47% over the number of Americans who were performing online banking in late 2002." In 2006, that number had increased: "Fully 43% of Internet users, or about 63 million American adults, bank online." See http://www.pewinternet.org/PPF/r/149/report_display.asp, accessed July 10, 2007, for more information.

65 and older, in another 10 years the people who are 55 to 65 and older will be much more technology-savvy. Thus, a forward-looking electronic service strategy should contend not only with the constituent population of today, but also with the likely constituent population of the future, which will inevitably be more technologically sophisticated. In addition, transaction and information services online have a significant cost benefit versus in-person or call-center alternatives.

The public is becoming accustomed to gaining access to personal banking information through bank or brokerage Web sites that present a smooth, seamless interface to a wide range of related services. Thus, banks, for example, now present to the public comprehensive Web sites through which customers can access information about their checking accounts and mortgage balances, while also viewing real-time stock market information, and can also access increasingly comprehensive ranges of other financial information and services. Some of the information accessible through such portals (for example, stock market data) is not owned by the portal maintainers themselves but is furnished as a convenient service. There seems to be a growing expectation that institutions such as financial services organizations will provide comprehensive access to data that their customers feel they need. This, too, is a characteristic of the world that the SSA will increasingly have to come to terms with. Such expectations are likely to hold, even though most people will deal with the SSA only infrequently.[2]

The SSA's many users are active participants in this world that is continuing to embrace such comprehensive 24/7-Internet-accessible services. Internet-connected computers are increasingly present in homes, and a growing majority of the public increasingly looks to these computers as vehicles for information and timely, convenient, online transactions. Although younger individuals may have been the first to grasp this new medium as a routine source of both information and service, older seg-

[2]The partnerships that enable the consolidation of financial information and services are examples of a "virtual value chain." The metaphor of the chain includes the notions of links, which the service provider may or may not control directly but may use for a fee or through partnership. In the "brick-and-mortar" world, the classic example of the value chain is the auto industry that flows from raw material providers and parts suppliers to the manufacturers to the distributors and dealerships that provide everything from parts to sales to service. Each link in the value chain does what it is best at. Typically, the customer does not deal with the individual links in the chain, but instead with whichever link is assigned to customer-facing issues. There are analogies in the e-business world; hence the term "virtual value chain," where the products being sold are information products and services instead of cars. The virtual value chain exists in e-government just as it does in e-business, with IRS e-file being one well-known example. S.H. Holden and P.D. Fletcher, "The Virtual Value Chain and E-Government Partnership: Non-Monetary Agreements in the IRS E-File Program," *International Journal of Public Administration* 28(7-8):643-664, 2005.

ments of the public are making similar changes to their expectations and habits (see Chapter 4). Search services such as those provided by Google and Yahoo! are increasingly used to locate information and service providers. Both public and private service providers are increasingly encouraging online access to their services: the public is becoming accustomed to shopping online, tracking the progress of parcel deliveries online, paying bills online, and renewing driver's licenses online.

One important reason for this shift in delivery channels is that service providers generally find that online transactions are less costly than in-person or telephone transactions.[3] This situation has changed significantly in the past 10 years, especially in the United States and to the advantage of online transactions.[4] Also, the 24/7 availability of such services makes them particularly convenient for customers and other users; convenience builds customer satisfaction and loyalty, both to the institution and to the online service channel, making it even more attractive for the service providers. Modern retail and institutional businesses generally view customer satisfaction as a primary metric, along with cost.

With respect to the SSA's mission requirements, the delivery of financial services has some compelling analogies to the delivery of benefits to citizens. For example:

- Customers expect their financial institutions to keep track of their accounts flawlessly and to handle transactions with complete privacy, security, and efficiency;
- Large financial institutions have transaction volumes comparable with those of the SSA (partially due to consolidation[5] in the bank industry); and
- A large proportion of customer transactions have moved online in the past 10 years.

[3]For instance, an Organisation for Economic Co-operation and Development (OECD) study reported that by the late 1990s, traditional branch banking customer service cost $1.08 per transaction, compared with $0.13 per transaction for Internet-based e-business. (That is, e-business provided an 88 percent cost savings per transaction.) That same study reported that telephone-based banking customer service cost $0.54 per transaction, indicating significant cost advantages to shifting service from phone to online. OECD, *The Economic and Social Impact of Electronic Commerce: Preliminary Findings and Research Agenda*, Paris, France: OECD, 1999, http://www.oecd.org/dataoecd/3/12/1944883.pdf, accessed June 20, 2007.

[4]With the convergence of voice and data and the emergence of Voice over Internet Protocol (VoIP) and speech recognition, the cost differences between call centers (for automated transactions) and the Internet are narrowing.

[5]The efficiencies provided by integrated electronic services have been substantial drivers for the consolidation that has taken place.

The following sections describe the state of the practice in the financial services industry and note parallels and potential lessons for the SSA and other agencies. While the committee is focused on the SSA, the lessons here are broadly applicable and should not be construed to imply that the SSA is alone in facing these challenges.[6]

TYPICAL ONLINE CAPABILITIES

Online services made available by banks encompass essentially anything except cash withdrawal and check or cash deposit. Routine banking transactions now increasingly occur online, leaving customer service representatives at call centers and branches to handle the more complex situations (and sales opportunities). For brokerages, online transactions can comprise over 90 percent of their total interactions with their customers. There are several categories of online capabilities, and while each plays a role in the delivery of services, they are at different points of maturity and adoption. Several of these capabilities are described below.

Product Information and Service Aggregation

Ideally, all products and services delivered by a financial institution are described in detail on the institution's Web site. This amount of information can be overwhelming, so in virtually all cases there are three ways in which information tends to be found and accessed: by looking for links corresponding to a need or precipitating event (for example, marriage); by looking for a particular product type (for example, loan account); and through a simple text search (for example, "cd rates"). All three mechanisms are widely used by customers, although search has increased in importance as Internet users have become habituated to the search box as the point of entry to content.

Over time, customers have grown to expect that their relationship with a single institution can be managed almost entirely online in a seamless fashion. Therefore, although retail banking, credit card, mortgage lending, and brokerage units (for example) are usually separate business units within a financial institution, customers expect them to be available in one place, with one log-in. Although most in the industry no longer think of this as "aggregation," it has required financial institutions to gather information from different systems, with different rules, and often from legally separate businesses, and to make them appear to be a single cohesive system that provides "one-stop service."

[6]Nor should the relevancies discussed here be taken to suggest that the SSA is not aware of these considerations.

Another form of aggregation is the gathering and organizing in one place of account information from multiple financial institutions. This aggregation can also be applied to nonfinancial accounts such as frequent-flyer miles, e-mail accounts, and so on. Intuit's Quicken as well as several financial institutions' Web-based aggregation services (such as Bank of America's My Portfolio and Fidelity's Full View) are examples. Another type of aggregation service that is being adopted is Bill Presentment, which assembles and organizes incoming bills to make them ready for payment. Coupled with online bill payment, this service eliminates the round-trip first-class mail circuit of receiving a paper bill and paying it by paper check.

Relevance for the SSA and Other Agencies

In order to design and deliver content through a Web site with state-of-the-practice functionality, an organization should conduct a thorough review of content and access to that content. What are the obvious high-level "types" of material or services that users might be looking for? The committee has not conducted a detailed analysis of this issue, but examples for the SSA might be "retirement benefits," "disability application," or "earnings data." Web pages, especially home pages, that are "flat" in structure, with a large number of disaggregated links, can appear cluttered, disorganized, and confusing to users. (See also the discussion in the section entitled "Financial Institution Web Site Design" below.) The public expectation that an individual will be able to find and access all of his or her benefit accounts in one place will continue to grow. Accordingly, a seamless presentation of any benefits or services to users that is facilitated by effective search capabilities—even if the management of those benefits crosses organizational lines—is preferred.

Account Management and Money Movement

Account-management and information activities include such tasks as looking up balances and terms. These activities once comprised the bulk of volume of online "transactions" for banks. Simply answering the question "How much money do I have?" was much simpler to do online for a segment of customers than to use the telephone (where inquiries of this kind are usually handled by an automated interactive voice response [IVR] system and can be quite cumbersome), an ATM, or a branch teller. Most account-management features also allow users to query balances or move funds among accounts or products after signing on to the Web site once. As an example, most users of online banking can log in to their financial institution's Web site and see information on their savings and checking accounts, their mortgage, credit cards, and possibly their retirement funds if they have chosen to consolidate their financial dealings with one firm.

Money-movement activities in financial institutions include account-to-account transfers (within the institution), bill payment, and account-to-account movement to other institutions. Of these, bill payment services have caused the most dramatic changes in customer behavior. U.S. banks have historically played only a limited role in consumer bill payment. While some activities, such as recurring automated payments for fixed amounts (such as mortgage payments) and telephone-based "pay anyone" services were available in the 1990s, there was very little adoption of these services. Online bill payment is now used by approximately half of online customers. Bill payment has fundamentally changed those consumers' interactions with their banks, both in terms of frequency of interaction and "stickiness"[7]—a "sticky" Web site has features that cause visitors to spend more time on the site and to return to the site. Thus, stickiness helps retain customers in the electronic service-delivery channel.

Brokerages and brokerage arms of banks offer every type of trading capability from buying and selling of simple securities and mutual funds to limit orders and option puts and calls. Trading online is today the standard way for individual investors to interact with the markets.

The ability for consumers to manage their accounts interactively online presents myriad security and access-control challenges. While this report's focus is not on security per se, it is a critical component of any electronic services strategy, as the SSA is appropriately aware. As financial institutions deploy and advance their own security strategies, there will likely be lessons for the SSA and other government agencies in how those institutions proceed.

Authentication for consumers on Web sites of financial institutions generally consists of a log-in and a password. Some of the banks with the highest rates of adoption for online services have used an authentication scheme that is already known to the customer—his or her ATM card number and personal identification number (PIN)—and required no extra enrollment step. The October 2005 guidance on authentication from the Federal Financial Institutions Examination Council has been changing this situation fundamentally; it requires multifactor authentication for banking transactions by the end of 2006.[8] Most banks have implemented some risk-based approaches to authentication (additional verification of

[7]See http://www.arraydev.com/commerce/JIBC/9908-03.htm, accessed June 20, 2007; and M. Khalifa, M. Limayem, and V. Liu, "Online Consumer Stickiness: A Longitudinal Study," *Journal of Global Information Management* 10(3):1-14, 2002.

[8]Federal Financial Institutions Examination Council, "Authentication in an Internet Banking Environment," FIL-103-2005, Oct. 12, 2005. See http://www.fdic.gov/news/news/financial/2005/fil10305.html, accessed June 20, 2007.

high-risk transactions such as funds transfer—for example, out-of-channel verification and/or stronger authentication) and employ layered authentication (for example, a minimum of identification [ID] and password, other knowledge-based checks, plus behavioral anomaly detection; two-factor authentication is also being introduced for the riskier interactions).

Another large issue here is better authentication of the financial institution's Web site and e-mails to the customer—which is currently a major vulnerability, given the prevalence of "phishing" scams.[9] The Financial Services Technology Consortium has launched the project "Authenticating the FI [Financial Institution] to the Consumer" to address this issue.[10]

Relevance for the SSA

Despite the SSA's early unsuccessful experience with making the Personal Earnings and Benefit Estimate Statement (PEBES, discussed briefly in Chapter 4) available online, if the SSA wishes to deliver state-of-the-practice electronic services, then online access to individuals' Social Security Statements of Earnings and other account information will be a required and basic function. Given the near ubiquity of such functionality in the financial services sector, as time goes on its lack will appear increasingly strange to a user base that is ever more technologically experienced. The implication is that users are accustomed to a high-value proposition in their online transactions that makes it more attractive to use electronic transactions than transactions by telephone or in person. If the SSA does not provide such a compelling set of services online to its users, they will remain loyal to the traditional service-delivery channels that they have become comfortable with in the past.

The SSA will never (and should not) remove the customer's ability to contact a human if desired; however, in order to attract and retain users in online channels, the SSA should strive to provide a consolidated view to users across program lines, even if the programs are run by different parts of the SSA organization—for example, a consolidated view of retirement benefits and disability benefits. Such provision may benefit from technology and data infrastructure that would support the capability for users

---

[9]"Phishing" is an attack that tricks a user into entering sensitive information (such as account numbers, log-in names, and passwords) at the wrong Web site, making it available to attackers. The most common example is an e-mail from a bank stating that there is an irregularity in payments or accounts that the recipient must attend to, and a link for the recipient to follow. But the underlying link takes the recipient to the attacker's Web site, which looks like the bank's legitimate site. The recipient follows the link and enters the information to access the account. The attacker can now also do so.

[10]See http://www.fstc.org/projects/current/authfi_home.2007.php, accessed June 20, 2007, for more information.

to authenticate themselves once to gain access to a variety of information and services across the SSA. (As more users access services over the Internet, "phishing" may also become a larger security risk for the SSA.)

Customer Service

In financial institutions, straightforward customer service activities such as opening or applying for accounts, ordering checks, changing one's address, and changing beneficiaries have moved online steadily. When third parties are involved, such as in ordering checks, and even in many cases when they are not, financial institutions had been slow to offer the option of "pre-filling" known customer information, making the online process more tedious. The more advanced sites now deliver that capability to authenticated customers and, as a result, provide more personalized services, thus making those services more convenient for users. With respect to opening accounts, banks and brokerages have been moving steadily away from simply making application forms available online toward offering fully interactive, "pre-filled" and streamlined application processes.

Relevance for the SSA and Other Agencies

Maintaining multiple channels for people to access services is important (apart from a very few online-only banks, financial services institutions maintain several use channels). Indeed, there may be lessons to be learned from the financial services sector with respect to state-of-the-art telephone triage and automation as well. Nevertheless, the state of the practice in retail financial institutions is to make available online all common services that are feasible from the customer's perspective. When the SSA is unable to provide a service electronically—for example, owing to regulation—clear explanations and alternative means of access are an important part of a comprehensive service-delivery strategy.

Where appropriate, a form or forms should be made available (not just as a printable blank form, but as a printable form that can be both filled in and submitted online) with detailed instructions on other needed materials and on how the applicant can complete the process. The state of the practice for online services today is that common information should only be provided once. For the SSA, this suggests that when a life event requires multiple forms for multiple programs, the common information should only be requested once. Pre-filling for authenticated users (such as current beneficiaries) should be the norm,[11] with clear explanations

[11]The committee understands that this capability is under development.

and alternatives provided for exceptional cases. This kind of functionality will have implications for the interfaces between the front-end electronic service sites and the back-end databases, processes, and infrastructure. (See Chapter 3 for more on technological considerations.)

In general, the maintenance of multiple channels should not preclude the transformation needed to provide services comparable with those of a large-scale financial services institution—such institutions maintain multiple channels themselves. As described in Chapter 4, e-government will eventually become simply "government" (and, accordingly, electronic services should simply be thought of as "services").

FINANCIAL INSTITUTION WEB SITE DESIGN

Web Design Principles

It is not accidental that the top financial sites are alike in many ways, and it is not due to one's copying another; this is an industry in which every participant expends great energy differentiating itself from its competitors. There are no hard-and-fast rules for the "best" design for a bank or brokerage site. However, a set of common principles has emerged, in addition to those standards and best practices codified by organizations such as the World Wide Web Consortium (W3C). These principles, relating to simplicity, navigation, and accessibility, are described below. Most of the companies' design processes incorporate significant user input and an understanding of appropriate usability mechanisms and techniques—this drives similarity in information architecture and navigation.

Relevance for the SSA and Other Agencies

Universal design principles should be applied to Web sites and usability testing should be made a routine part of the design process.[12] Before sites face actual users, it is imperative to adopt user-centered design methods in which the needs and constraints of users are taken into account at all stages: that is, in design, development, deployment, requirements analysis from user perspectives, user participation in the design process, and user testing.

In addition, careful analysis about what kinds of services are most analogous to those offered by financial services institutions (and what are not) will be needed. Where sufficient similarity is found, not everything

[12]See, for example, B. Shneiderman and C. Plaisant, *Designing the User Interface*, 4th ed., Addison Wesley, Boston, 2005; or S. Krug, *Don't Make Me Think: A Common Sense Approach to Web Usability*, 2nd ed., New Riders Press, 2005.

needs to be tested; many of the most common practices in industry can be assumed to be the current best practices. Care must be taken, though, not to assume that the customer bases are necessarily similar; demographics and user expectations of the SSA user base may differ in subtle ways from those of major financial institutions, requiring attention when designing interfaces. Appropriate attention to standards and best practices articulated by Web standards organizations such as the W3C may also be helpful. Of particular importance is the need to avoid the assumption that everyone uses or has access to one or a small number of browsers. In general, as more is learned about how people interact with the Web and as the ways that people interact with Web sites change, the standards and best practices for the design of effective and usable Web sites will change. The SSA Web site's design will need to evolve as the best practices and standards evolve.

Navigation and Simplicity

Any effective financial services site will have uncluttered pages with clear emphasis on the most common and natural transaction at each step. Achieving effective simplicity can be a more subtle challenge than one might think at first. In the case of basic Web design principles, counting the number of clicks to a transaction was once the core metric applied. This metric led to much "busier" pages and more choices at each step. This way of measuring simplicity has been relegated to a secondary measure, primarily for the following reasons:

- Considerably-more-complex pages are tolerable for users whose Internet access speeds have improved (owing to increasing penetration of broadband access and high-speed Internet access from the workplace, libraries, or other institutions);
- New technologies that allow pages to be dynamically updated eliminate the need for a complete page refresh; and
- Users have become more sophisticated.

Simplicity also applies to visual clutter; if the eyes cannot find what they are looking for among things too similar to each other, then the user is likely to fail. Having good graphical design, putting similar things near each other, using wording that is clear—all lead to successful designs. An additional simplifying feature, from the user's perspective, is for the back end to be flexible and robust in terms of how it handles data. Many Web sites distinguish themselves by both being maximally flexible in the syntax that they accept (for instance, telephone numbers may be entered with or without blanks, dots, or hyphens) and maximally consistent in the syntax that they produce (for instance, always exhibiting telephone numbers consistently broken down into area code, exchange, and subscriber number separated by hyphens). An emphasis on robust data handling and presentation—accepting anything reasonable but always producing a canonical form—can achieve a huge improvement in data quality and in user experience while the user is entering data.

A navigational norm has emerged, consistent with many other commercial sites. If a single domain name serves multiple natural audiences, the navigation bar at the top allows users to indicate what type of customer they are and/or what services they are interested in (see the first two examples in Figure 2.1). Alternatively, they may navigate by type of product or service they are seeking (see the second two examples in Figure 2.1).

Common elements on the sorts of sites illustrated in Figure 2.1 are "search" and "log-in." The log-in function is the gateway between the authenticated and unauthenticated parts of the site. For virtually all modern financial sites, the user experience (look and feel, navigation) is the same whether one is on the authenticated or unauthenticated part of the

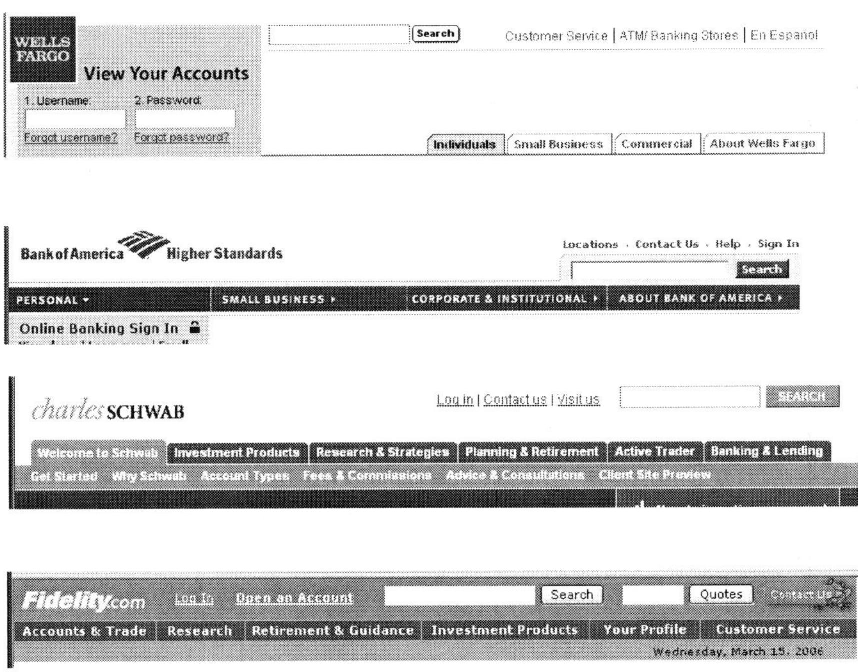

FIGURE 2.1 Examples of typical commercial navigation styles.

site. Although the technology behind sites can be quite different, that distinction is designed to be transparent to the user.

There is less consistency than ever in the use of underlining for links. This was a virtual requirement in good design 10 years ago, but that is no longer the case. Through extended use of the Internet, along with acclimatization to interaction online, users have learned to recognize other cues, now expecting nearly everything except paragraphs of text to be a link. The use of underlining for link-intensive pages or sections of pages causes visual clutter and reduces readability.

Relevance for the SSA and Other Agencies

As a simplifying measure and as a way to emphasize a service orientation, content on the Web site regarding the agency itself (as opposed to the services that it offers) might be placed in a section subordinate to material that emphasizes services at the top level. A navigation bar should be composed of "parallel" items in some dimension. "Search" should have a text-input box. It is particularly important to understand the primary modes of use represented by those who will visit this Web site and to design labels suited to those individuals: for example, "Newly Retired," "Checking Benefits," or "Are You Ready to Retire?"[13]

Accessibility

All well-designed financial services sites are developed to be "accessible" to the visually impaired, in the technical sense of their information or content pages being W3C accessibility compliant[14] and thus navigable by a screen reader (such as Job Access with Speech, or JAWS[15]). That is more challenging for the transactional portions of sites, but progress has been made there in the past 5 years.

Relevance for the SSA and Other Agencies

The SSA Web site, as well as other government Web sites, is expected to be at least Section 508 compliant; the SSA should ensure that all internal

---

[13]Among government portals, FirstGov.gov offered an example of a reasonably effective use of usage-mode-oriented parallelism in the way that it presents its options.

[14]See World Wide Web Consortium, Web Accessibility Initiative, "Web Content Accessibility Guidelines 2.0 (WCAG 2.0) Working Draft," Nov. 23, 2005, at http://www.w3.org/WAI/, accessed June 20, 2007.

[15]JAWS (Job Access with Speech) is a screen reader for the visually impaired. See http://www.freedomscientific.com/fs_products/JAWS_HQ.asp, accessed June 20, 2007.

and external applications are also compliant with Section 508[16] and, to the extent possible, with other accessibility guidelines (such as those developed under the auspices of the W3C[17]). The SSA has a particular interest in accessibility given that it serves the disabled community explicitly. As more sophisticated electronic services are made available, maintaining accessibility will be important going forward, especially given the broad, diverse, and heterogeneous user base of the SSA.

CUSTOMERS AND USERS OF ONLINE SERVICES

Over 90 percent of U.S. households maintain a "transaction account" (such as a checking account, bill-paying account, or share draft account) with a bank, thrift institution, or credit union.[18] Nearly all of these institutions offer online capabilities to their customers, and almost half of their customers make use of some or all of these services.[19]

Getting to these levels of penetration has taken 10 years of sustained and focused effort across the industry.[20] Banking customers in the United States, even those most devoted to the online channel, remain largely "multichannel," with consumers making use at the very least of a bank's

[16]Per the Web site http://www.section508.gov (accessed June 14, 2007), "In 1998, Congress amended the Rehabilitation Act to require Federal agencies to make their electronic and information technology accessible to people with disabilities. Inaccessible technology interferes with an individual's ability to obtain and use information quickly and easily. Section 508 was enacted to eliminate barriers in information technology, to make available new opportunities for people with disabilities, and to encourage development of technologies that will help achieve these goals. The law applies to all Federal agencies when they develop, procure, maintain, or use electronic and information technology. Under Section 508 . . . agencies must give disabled employees and members of the public access to information that is comparable to the access available to others."

[17]See the Web Accessibility Initiative at http://www.w3.org/WAI/, accessed June 20, 2007.

[18]In 2004, the Federal Reserve Board found that 91.3 percent of families hold transaction accounts. See Brian K. Bucke, Arthur B. Kennickell, and Kevin B. Moore, "Recent Changes in U.S. Family Finances: Evidence from the 2001 and 2004 Survey of Consumer Finances," *Federal Reserve Bulletin*, Vol. 92, February 2006, pp. A1-A38 (Table 5), available at http://www.federalreserve.gov/pubs/oss/oss2/2004/bull0206.pdf, accessed July 10, 2007.

[19]Ted Schadler, Charles S. Golvin with Jed Kolko, Sally M. Cohen, and Tenley McHarg, "The State of Consumers and Technology: Benchmark 2005" in Forrester, *Consumer Technographics North America*, available at http://www.forrester.com/go?docid=36987, accessed June 14, 2007.

[20]The experiments in so-called online banks that allow only online interaction peaked in the bubble years of 1999-2001. Of these, Wingspan and First Security were ultimately unsuccessful, for example. Although some, such as NetBank, have survived, their impact on the industry has thus far been negligible in the United States. There are, however, some special cases, such as institutions offering above-market certificate of deposit rates, where online-only institutions (such as ING) are beginning to gather substantial deposits.

ATMs for cash and check-deposit needs, and more generally remaining at least occasional users of branches and call centers.

The retail brokerage industry serves a smaller client base. The majority of brokerage accounts are with "lower cost" brokers, formerly known as "discount" brokers, such as Charles Schwab, Fidelity, E*Trade, and Ameritrade. Nearly all customers of these brokers, as well as a majority of clients of the traditional "wire houses" (such as Merrill Lynch) have sophisticated online capabilities. Trading online has become the default execution method; adoption of online services has transformed the retail brokerage industry.

Relevance for the SSA and Other Agencies

Much of the financial services industry, which previously relied heavily on paper-based transactions for customer interactions, has shifted large proportions of its interactions with consumers online. Certain market segments, primarily those early adopters with more income, education, and computer-savvy, are both comfortable and experienced with e-commerce. As a result, they should be comfortable with e-government transactions—even those involving their benefits—because they have already completed analogous transactions online with private-sector financial institutions. Looking ahead to the kinds of services that might be expected in the future, these same users might also value being able to understand their complete financial picture online if the financial services industry had the capacity to display information from the SSA as part of users' financial profiles without rekeying the data—albeit such a service raises the usual privacy and security considerations.

BUSINESS INCENTIVES FOR ONLINE SERVICES

Banks and brokerages have made significant investments in developing their online service channels. One motivation has been that the cost per transaction differs greatly among service-delivery channels. The business case for doing so is composed of multiple elements: customer satisfaction, customer retention, acquisition of new customers, cross-sales of other products to existing customers, and the economic trade-offs between cost reduction and the cost of delivering online services.

Customer Satisfaction

A solid set of well-designed online capabilities improves customer satisfaction. In the past 10 years there has been ever-increasing focus on customer satisfaction as the financial services industry has consolidated and personal relationships with local bankers have been replaced by interactions with megabanks. In the leading institutions, customer satisfaction is carefully measured and tracked. These banks can therefore measure the impact of online capabilities through matched samples of online customers versus customers using other service channels (though finding that control group for predictive purposes becomes more difficult as online usage becomes increasingly ubiquitous). Although the banks' findings are proprietary, in general the results are thought to be very positive, with customers who conduct much of their business online showing significant improvements over time in their overall view of the banks.

Relevance for the SSA and Other Agencies

Appropriate measures of customer satisfaction require data from individuals using all of the agency's service channels. This means that a representative sample of all individuals who interact with the agency, whether by phone, in an office, or online, should be consistently surveyed on a few simple measures of service satisfaction. One important observation is that the effectiveness of electronic services should be determined by measuring *relative* overall satisfaction of the users of online services *as compared with* the satisfaction of the remaining population of individuals using phone or office interactions (as opposed to measuring "satisfaction with online service" by itself). Note that the topic of customer satisfaction addressed in this subsection is only one of the many reasons for encouraging the use of electronic services—others include efficiency, accuracy and improved error rates, and convenience.[21] Monitoring customer satisfaction as part of an overall metrics-based approach (see Chapter 4) to service provision can aid in encouraging the use of electronic services and in retaining users for that channel.

Customer Retention, New Customer Acquisition, and Marketing

Most Americans inevitably turn to the SSA at key points in their lives. Therefore, the SSA does not need to "acquire" new customers or "retain" them in exactly same way that a commercial concern would seek to attract new customers and keep its customers from switching to a competitor. However, the advantages that can be gained by the SSA and the public

[21] A 2006 OECD paper proposing an inventory of business case indicators for e-government initiatives noted both reductions in benefits mispayments and in savings efficiency from both time savings for public servants and reduced error rates and rework. OECD E-Government Project, *Proposal for Work on an Inventory of E-Government Business Case Indicators,* Feb. 6, 2006, available at http://webdomino1.oecd.org/COMNET/PUM/egovproweb.nsf/viewHtml/index/$FILE/GOV.PGC.EGOV.2006.3.doc, accessed January 4, 2007.

through wider use of electronic services do create motivation for the SSA to try to acquire and retain users in electronic channels of service delivery. Toward that end, the learning curve of commercial financial institutions can be informative for the agency.

The effect of online services on retaining existing customers has been the most compelling element of the case for introducing, expanding, and encouraging the use of these services. The rule of thumb in the banking industry is that it costs two to four times as much to acquire a new customer as it does to retain an existing one; in addition, the revenue generated by a new customer is often significantly less than the one they "replace." Retention, therefore, is a key operational metric in any financial institution. The search for "sticky" products or services has been a long one, and none has succeeded like online access and—particularly, and quite dramatically—like online bill payment.

In contrast, online service offerings in retail banking have not had a dramatic impact on acquiring new customers. This is due to two countervailing forces. In the early years of online banking (1996 through 1999), there was significant differentiation among banks regarding their online offerings, but the demand for those offerings was quite low. In recent years, demand has increased dramatically, but now virtually every bank offers online access. Certainly the quality of the online services varies substantially, but that is apparent only to actual users of the services and much less so to those "shopping" for a bank. By contrast, in the retail brokerage industry, the situation is dramatically different. New entrants (with E*Trade and Ameritrade being the principal survivors) and more nimble "discount" brokers used their online expertise to draw in a whole class of new customers, as well as converting customers from the traditional "full service" wire houses.

The marketing of online services has taken multiple forms. For retail brokerages, as the bulk of their business has become e-business, virtually all of their marketing—whether print, television, or online—has placed their electronic services in the center of communications. For retail banks, marketing online services is one of the arrows in the communications quiver. It has become standard practice for banks to actively introduce and promote their online capabilities to new customers at the time of account opening, even if that takes place in the branch. For both the banking and the brokerage industries, one of the most effective means to convince customers to migrate to the online channel is to inform them politely at all other points of contact—for example, on the telephone or in person—that the particular transaction they are undertaking could be done more easily online (some banks even offer kiosks in their walk-in lobbies that facilitate online access to customer accounts). Brokerages now conduct the bulk of their transactions online; this may be partially due to the fact that the demographics of users of brokerages are even more conducive to e-commerce adoption than are the demographics of users of banks.

Relevance for the SSA and Other Agencies

Although the SSA may not need to "acquire" new customers, it does need to attract them to—and retain them in—electronic channels. As noted earlier, having users' needs met in the electronic services channel frees up resources for other users who need more-personalized service or responsiveness from other channels. This "stickiness" (retention) matters because the increasing use of an effective electronic channel will help the SSA keep costs down; it will mitigate shortages of in-person staff to handle the expected workload as the baby boom generation ages, becomes disability-prone, and retires; and it will likely increase public satisfaction with SSA information and services.

Public expectations for financial institutions with respect to functionality and availability are likely to translate to public expectations for electronic information and services from government agencies. At the same time, shortcomings will likely frustrate people and keep them in the more costly (and strained) delivery channels such as telephone calls and in-person visits. In addition, careful consideration of demographics and of particular market segments that are using electronic services at a high rate might lead the SSA to consider marketing to those segments very aggressively. There may also be lessons in how other government agencies have reached out to particular customer segments (see the discussion of the Internal Revenue Service in Chapter 4). Most importantly, the SSA needs to be able to promote the relative benefits of electronic services over the competing service channels in ways that are meaningful to users. The marketing needs to answer the question of, "What's in it for me?" in a clear and compelling way.

Cross-Selling

Selling additional products or services to existing customers has generally been more successful through the online channel for financial services institutions. As compared with physical channels, the online experience more readily allows customers to appreciate the convenience of having "everything in one place"—even at a branch, a loan application might take place in a physical space different from where one makes a deposit; the online interface can hide even that sort of process separation. This drives the desire on the part of customers to consolidate their financial activities. In addition, a user of online services can be marketed to in a highly targeted way, at very low cost.

Relevance for the SSA and Other Agencies

Successful electronic services induce use of other available electronic services, potentially producing more effective and efficient service delivery overall. This improvement includes a reduction in data-entry error rates and the resultant costs of reworking to correct errors. Because of the relatively higher customer satisfaction scores for electronic services compared with telephone and paper processes, it stands to reason that a satisfied user of one e-government service will be more likely to seek out similar services to avoid long phone queues or paper-based cycle times.

Cost Trade-offs

Marginal costs of online transactions are very close to zero, and in any case much lower than any other means of delivery such as in person or by telephone. The fixed costs of building and maintaining online systems are significant, but the payback period—at least for the initial outlay—should be relatively short, even assuming that transaction volume remains the same. But, indeed, transaction volumes have tended to increase, as customers with online access have been observed to increase their overall use of services quite significantly. As these services have marginal costs that are close to zero, the costs incurred due to increased volume are outweighed by the savings due to the reduction in need for call center staff and IVR calls.

Relevance for the SSA and Other Agencies

As the SSA seeks to improve efficiency and migrate users to electronic channels, systematic cost tracking can help assess where additional efforts would prove fruitful. Such cost tracking should not be just at the aggregate level but instead should examine cost per transaction through a methodology such as activity-based costing. The effects of each service offered online with respect to call and office-visit volume could be systematically tracked as part of an overall effort to monitor appropriate metrics. Although the attribution of changes in call or office-visit volumes to the use of particular online services can be difficult, a standard and consistent methodology should be developed and agreed on. Items to be measured might include things such as costs per transaction, satisfaction, error rates, cycle times, repeat users by delivery channel, and so on. All of this should be done with a goal of maximizing value to users (taxpayers) and minimizing cost.

ORGANIZATIONAL STRUCTURES FOR ONLINE SERVICES

When considering e-business and electronic services generally, it appears that management and organizational structures for such services have had to proceed through three phases in the largest and most successful companies. In the first phase, which for most banks was approximately 1996 through 1998, there were multiple "e-groups" emerging in the organization. The primary focus for these services was often in marketing and communications organizations, as the chief use of Web sites was to market services rather than to provide them. In the second phase (approximately 1999 through 2002) "e-business" was paramount. Centralized, autonomous e-commerce lines of business—reporting very high in the organization—were formed with virtually end-to-end control of the e-channel. This seems, in retrospect, an almost necessary step to developing a sophisticated strategy, infrastructure, and set of policies for electronic service provision.

In the third phase, the current state of maturity (approximately 2002 to the present), the underlying approach appears to be "e-business is business" (see Chapter 4 for this committee's presentation of the idea that now, or very soon, "e-government is government"). Although some of the largest and most successful institutions have kept a large e-group at the core, virtually every line of business and function has developed expertise to leverage the core assets. For example, organizational roles have tended to evolve along the following lines:

- *Centralized e-group:* Strategy development, funding, navigation and information architecture, content ownership assignment, look and feel, policy development and enforcement, business requirements, storyboarding, wireframing, page design, sometimes front-end development;
- *Marketing and communications functions:* Content for the "About" section, sign-off on the overall look and feel to fit the brand architecture;
- *IT function:* Infrastructure, back-end and middleware development, sometimes front-end development; and
- *Lines of business:* Content, business requirements for transaction services jointly with the core group.

Relevance for the SSA and Other Agencies

The SSA's management and organizational structures for electronic services and e-government have not yet passed through the second phase described above. To track well with the maturing electronic services knowledge in the financial industry, in the second phase the development and management of electronic services would be centralized and elevated in the organization.[22] See Chapter 4 for more on this issue.

SUMMARY

In the future, the Social Security Administration will increasingly be viewed as a financial institution whose services are needed, and hence whose services will be expected to be available, on a 24/7 basis. Taking advantage of the experience of commercial financial institutions can help the SSA as it orients itself technologically and culturally toward weathering the oncoming storm of increasing workload, workforce transition, and changing public expectations. The commercial financial industry's history of developing and marketing online services can provide the SSA with relevant experiences in the areas of adoption patterns, economic incentives, typical capabilities, Web site design, and organizational structure. Incorporating these lessons will require a strategic focus on electronic information and service delivery, metrics-guided improvement, and process transformation.

Finding: The experiences of large-scale financial institutions in transitioning to the provision of electronic services are instructive in considering the challenges faced by the SSA in formulating its medium- and long-term electronic services strategy.

Recommendation: The SSA should carefully consider the ways in which the experiences and approaches of large-scale financial institutions—including state-of-the-practice electronic information and service delivery, metrics-guided improvement, and process transformation, among other approaches and solutions—might be relevant to the kinds of services that the agency is providing or may provide in the future.

[22]Julianne Mahler and Priscilla Regan have documented how federal agencies' management of their Web resources has evolved over time, tending toward more-centralized control mechanisms. Many agencies have struggled with issues involving the degree of centralization in policy making and the locus of responsibility within the agency. See J. Mahler and P. Regan, "The Evolution of Web Governance in the Federal Government," *International Journal of Electronic Government Research* 2(1):21-35, 2006.

3

The Social Security Administration's Information Technology— Present and Future

As the Social Security Administration's (SSA's) workload increases, technological change continues and brings with it a corresponding increase in the online service expectations of at least some segments of its clientele. Unlike earlier retirees, the baby boom generation is more computer-literate and Internet-savvy and is likely to retain these traits after reaching retirement age.[1] Thus, it should be expected that the SSA's various beneficiaries and other user communities will continue to be adept at keeping pace with these technological changes. Broader and more systematic attention to technical and social trends will be warranted. Further, to achieve a forward-looking electronic services strategy in any organization requires an understanding of the organization's current technological capacities coupled with an understanding of trends in important and relevant information technology (IT) and user expectations. This chapter provides a high-level assessment and set of impressions regarding the SSA's current IT infrastructure, its database technology and conversion strategy in particular, external technological trends, a brief overview of user expectations and projected demographics, and how they all can affect prospects for effective electronic services, now and in the future.

[1]See the discussion of Pew Internet and American Life Project survey data later in this chapter.

INFORMATION TECHNOLOGY INFRASTRUCTURE

The IT infrastructure of an organization encompasses hardware, software, databases, applications (including Web services) along with security processes and, for the purposes of this report, software-development policies and practices. This section briefly describes the committee's assessment of various of these parts of the SSA's IT infrastructure. It focuses particularly, however, on the underlying databases and database architecture, as these are especially key components of any effective electronic services strategy and application suite (see Box 3.1 for more on the centrality of databases to electronic services provision).[2]

Hardware Infrastructure

The SSA organizes its hardware and IT infrastructure[3] in three tiers: local/departmental, remote operations control centers (ROCCs), and headquarters. There are numerous local offices; six ROCCs located in Pennsylvania, New York, Illinois, Missouri, California, and Alabama; and one headquarters facility, the National Computer Center (NCC), located in Baltimore, Maryland.

The hardware in the local offices consists primarily of personal computers (PCs) running the Microsoft Windows operating system, plus a local file server, all connected using standard local area network (LAN) technology. Desktop machines are upgraded regularly on a 3- or 4-year cycle. In addition to access to the local servers, local PCs also have access to the servers at the ROCCs through virtual private network (VPN) connections. The ROCCs run standard Unix servers from a mix of vendors and again are connected using standard LAN technology.

The bulk of the SSA programmatic applications run at the NCC. The NCC hardware configuration is typical of a large financial services organization. It consists of a variety of hardware configurations, including midsized servers running both Windows and Unix with their own storage volumes, six IBM Parallel Sysplex mainframe systems, and a large data-storage farm implemented using storage area network (SAN) technology from EMC Corporation. Most of the programmatic data sets reside on the

---

[2]Note that this overview is of necessity brief and based on comparatively small amounts of data and input. A comprehensive assessment of such a large organization's IT infrastructure was outside the scope of this committee's activities; the committee tried to focus particularly on the capabilities and functionalities related to electronic services provision.

[3]Factual details about the SSA's IT infrastructure in this section are from Social Security Administration, *Information Resources Management Strategic Plan (2005)*, especially pp. 85-224 (hereafter cited as Social Security Administration, *Information Resources Management Strategic Plan (2005)*), available at http://www.socialsecurity.gov/irm/IRM_2005.pdf, accessed June 20, 2007.

BOX 3.1
Modern Database Technology Is Critical to the SSA

The rapid growth in the development of large-scale electronic services has been strongly supported by rapidly growing understanding of the anatomy of such services and by advances in supporting the development and deployment of their component parts. The data-management component is the most critical component—online services 24 hours a day, 7 days a week (24/7) are as dependent on the data-management component as an automobile is on its engine.

Virtually any information technology (IT) application, whether it is in the public or the private sector, is about data. In the case of the Social Security Administration (SSA), the central data comprise the information that the SSA has about its core constituents—essentially every person and business in the country. That is at the very core of the agency's mission—it keeps track of everyone: who they are, who their families are, where they live, what they have earned each year, what they or their employers on their behalf have paid to the SSA. This is all data. It is what drives everything that the SSA does. All of the SSA processes and services are centered on those data—they read the data, they record changes to it, they note what is happening to it, and so on. The data are the model that the SSA has of everyone—they are its computerized model of the world that the SSA exists to manage.

Electronic services must rely pivotally on appropriate database services. Requested services may simply require access to up-to-the-minute information, in which case this must be available from an archival data store that is capable of receiving and processing updated information in real time. The privacy of this information, however, must be carefully guarded, and its security from loss or tampering must also be ensured. These demanding requirements must be supported by the archival data-storage component. The services requested may also entail modification of archival information (for example, adding in the latest Social Security employer deposit, or logging the payment of a monthly benefit). Here too, the data-storage component is relied on for assuring the provision of the appropri-

EMC storage arrays so that they are available from any of the mainframes. All of the machines are interconnected using standard LAN technology, which is protected from outside access by a firewall.

The SSA's continued dependence on Customer Information Control System (CICS), Virtual Storage Access Method (VSAM), and the crucial Master Data Access Method (MADAM) software currently restricts the SSA to IBM mainframes for the bulk of its programmatic data. As discussed below, migrating these data to a relational database system would offer the SSA an opportunity to consider hardware alternatives with more performance at a much lower cost. For example, since the mid-1980s, the software company Teradata has provided the retail and banking sectors with scalable database technology based on the use of scalable servers constructed from commodity hardware components. In 2004, Wal-Mart had

ate capabilities. Thus, the data-storage component must be capable of immediate updating as well. The challenges entailed in developing such data stores clearly are considerable: they must be incorruptible, persistent in perpetuity, even in the face of attempts to corrupt them, and must also support being able to be updated in real time. Considerable research and commercial development, however, have led to the availability of such databases.

Support for 24/7 access to electronic services requires that all components be capable of providing appropriate support. Thus, clearly, the services promised by an inviting user interface can only be provided if an appropriately powerful database is a component of the application. Today, though, the SSA is using a decades-old database system. The Master Data Access Method (MADAM) is based on a 40-year-old proprietary record-based file/index system that ran on IBM 360s and still runs on their modern mainframe hardware. The fact that the SSA's equivalent of the "corporate jewels" are stored using decades-old technology is alarming and needs to change.

As Chapter 3 in this report points out, modern database technology can readily manage the amount of data that the SSA has. Moreover, the number of IT professionals having the background and skills to maintain the current MADAM database and surrounding applications is very small, and dwindling. Few people today know how to deal with mainframes and the kind of software that MADAM is based on. In the relatively near future, there may be none. By contrast, Structured Query Language (SQL) developers are ubiquitous. The SSA needs to move forward quickly in this area so that it can develop new services and processes using the many skilled developers and system administrators and performance tuners available in the modern job market. Otherwise it will continue to operate in the "old days" of multi-hour batch updates and backups, fragile software that few know how to tune or manage, and so on. The SSA will not be able to provide effective 24/7 services that way, nor will it be able to add new data to support new services easily. Modernizing the underlying databases that support the SSA's activities is critical to the agency's effectiveness in the realm of electronic services.

almost 500 terabytes (TB) of operational data stored on its 1,000 processor Teradata configuration.[4] Other scalable database alternatives include IBM's relational database system product DB2/PE (Parallel Edition), Oracle 10g, and products from vendors such as Netezza and Datallegro. The adoption of scalable database technology and clusters of commodity processors to handle even the very largest database tasks is expected to accelerate as databases of all forms and types continue to grow. Exploration of the use

[4]Constance L. Hays, "What Wal-Mart Knows About Customers' Habits," *New York Times* Business Section, Nov. 14, 2004, available at http://www.nytimes.com/2004/11/14/business/yourmoney/14wal.html, accessed June 14, 2007; and Charles Babcock, "Data, Data Everywhere," *Information Week,* Jan. 9, 2006, available at http://www.informationweek.com/story/showArticle.jhtml?articleID=175801775, accessed June 20, 2007.

of this sort of technology as part of a migration plan from MADAM to relational database technology would likely be fruitful.

Redundancy and backups are of course essential to data integrity and availability—and ultimately to business continuity. One means for meeting this need is through the use of data replication to provide the capability to switch over automatically to spare components that provide for continuity of operations. In general, the reason that such replication is important is that, although backup tapes are kept off-site in a secure location to prevent significant data loss, without replication a major disaster at the primary site could leave the SSA unable to access its master database until a replacement facility was up and running.

In this study, the committee was unable to obtain information from the SSA regarding data replication and failover plans for handling disaster recovery owing to security sensitivities. Therefore, rather than evaluating the agency's current or planned capabilities in this regard, the committee notes that various approaches to addressing this challenge are possible and are supported by modern database systems and architectures. However, in the committee's view, realizing a modern data-replication and disaster recovery strategy could be significantly hindered by the SSA's use of decades-old database technology for storing the majority of its programmatic data (see the discussion of MADAM in the following sections). The antiquated MADAM database system and technology are difficult to replicate using current technology (see below), and the custom modifications that would be needed to do so would be very costly and would require scarce expertise.

Software Infrastructure

Each of the three SSA organizational tiers runs slightly different software.[5] At the local/departmental level, both the desktop PCs and the servers run Windows NT and the usual suite of Microsoft productivity tools. Microsoft Access is run on the desktop PCs, and the local/departmental servers run Structured Query Language (SQL) Server. As is typical for large organizations, the desktops are used for running a broad spectrum of applications, including administrative applications (for example, payroll and travel); management information applications (for example, report generation); and programmatic applications (for example, data entry). Although the applications are currently a mix of client-server

[5]Factual details about the SSA's IT infrastructure in this subsection are drawn from Social Security Administration, *Information Resources Management Strategic Plan (2005)*, especially pp. 85-224; available at http://www.socialsecurity.gov/irm/IRM_2005.pdf, accessed June 20, 2007.

and Web-based applications, the SSA is in the process of replacing most or all of the client-server applications with Web-based versions. Both types of applications are used to communicate with servers located at all three tiers: local, ROCCs, and NCC. The servers at the ROCCs are a combination of Unix and Windows NT servers. Both Oracle and SQL Server database systems are used at this tier. The NCC hosts servers running Windows NT, Unix, and IBM's z/OS operating system. SQL Server is run on the Windows servers, Oracle runs on the Unix servers, and a combination of DB2, Computer Associates Integrated Database Management System (CA-IDMS), MADAM, and VSAM run on the IBM Parallel Sysplex servers.

The databases housed on the servers at the NCC can be roughly divided into three categories: programmatic, administrative, and management information (MI). The programmatic databases are organized into separate operational data stores; they include Title II (administering disability, old age, and survivor benefits); Title XVI (administering Supplemental Security Income [SSI]); Disability (determination, control, and tracking); Earnings (recording of annual wage reporting by employers and benefit reports); and Enumeration (allocation and verification of Social Security numbers [SSNs]). The administrative database is a separate operational data store that includes information on SSA employees, facilities, and finances. Each of these operational data stores feeds abstracted and aggregated information into a data warehouse that is used to support management decision-making processes based on both current and historical data. The schemas for each of these databases are consolidated in a metadata database. See Figure 3.1 for a graphical depiction of the SSA's Management Information Database Architecture.

The MI data warehouse is hosted using Oracle running on the Unix servers. The administrative databases and associated applications are currently in the process of being migrated to a combination of DB2 running on one of the SSA mainframes and Oracle running on a Unix server.

With the exception of the Disability database, which is housed on DB2 on a mainframe, the primary programmatic databases including Title II, Title XVI, Earnings, and Enumeration are stored on a SAN and manipulated using a combination of CA-IDMS, MADAM, and IBM's VSAM.

While the names DB2 and Oracle will likely be familiar to the casual reader, it is highly unlikely that CA-IDMS, MADAM, and VSAM will be. CA-IDMS is an implementation of the Conference on Data Systems Languages (CODASYL) data model. The company that actually built IDMS (Cullinet) went out of business in 1989, at which time Computer Associates acquired the rights to Cullinet's system and its customers. The CODASYL data model was popular in the early to mid-1970s, but by approximately 1980 most experts in the data-management field had

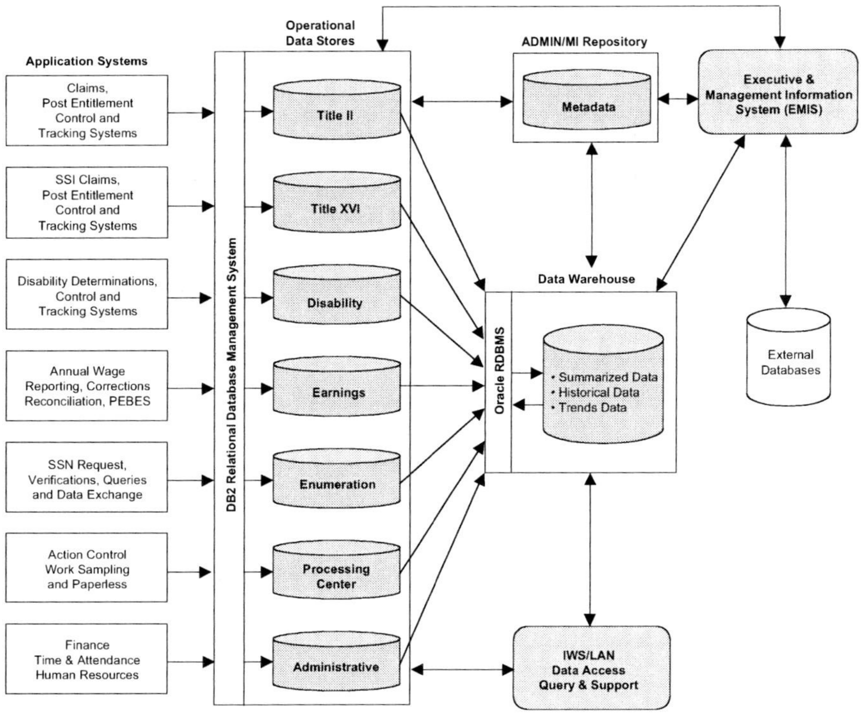

FIGURE 3.1 The Social Security Administration's Management Information Database Architecture.
NOTE: SSI, Supplemental Security Income; PEBES, Personal Earnings and Benefit Estimate Statement; SSN, Social Security number; ADMIN/MI, administrative and management information; RDBMS, relational database management system; IWS/LAN, intelligent workstation/local area network.
SOURCE: Social Security Administration, "SSA Data Architecture," Version 1.0, Dec. 9, 2002, p. 22. (Courtesy of the Social Security Administration.)

concluded that the relational data model was superior.[6] In addition, CODASYL database systems are almost exclusively accessed by applica-

---

[6]For a timeline history of databases, see International Federation for Information Processing Working Group 9.7, "A Brief History of Database Systems," in *History of Computing,* last modified Dec. 5, 2004, available at http://www.comphist.org/computing_history/new_page_9.htm, accessed June 20, 2007. For discussion of the foundations of database systems, see Raghu Ramakrishnan and Johannes Gehrke, *Database Management Systems*, New York: McGraw-Hill Higher Education, 2002, especially Ch. 3 (hereafter cited as R. Ramakrishnan and J. Gehrke, *Database Management Systems*). A mid-1970s comparison of the CODASYL and relational approaches is presented in Ann S. Michaels, Benjamin Mittman, and C. Robert Carlson, "A Comparison of the Relational and CODASYL Approaches to Data-Base Management," *ACM Computing Surveys* 8(1):125-151, March 1976.

tions written in Common Business-Oriented Language (COBOL), which, as the oldest business-oriented programming language in the history of computing, is now generally considered to be obsolete, understood by only an increasingly small fraction of the practitioner community.

One drawback is that applications written in COBOL/CODASYL are more cumbersome to maintain compared with those implemented using modern technology. An application written in COBOL against a CODASYL database management system (DBMS) such as CA-IDMS is typically 10 to 100 times longer (in lines of application code) than an equivalent application written in C++ or Java against a relational database system. The main reason for this difference is that the database requests against a relational database system are written in SQL, the declarative data manipulation language that is the mainstay of relational databases and modern IT organizations. Simply put, SQL allows applications developers to request data by saying "what" they want, not "how" they want the system to obtain the data. The underlying relational DBMS optimizes the application's request based on how the data are physically organized and indexed and what the current data statistics look like. This provides orders-of-magnitude code reduction and productivity gains relative to CODASYL database applications.

CODASYL applications are extremely difficult to write, debug, and maintain in comparison with today's mainstream data-centric application-development technologies. The SSA's dependence on CODASYL and COBOL for MADAM represents a severe handicap because of these difficulties in developing, refining, and maintaining the demanding applications needed to meet the current requirements of SSA's diverse constituencies.

Another drawback is that COBOL itself will eventually become a "dead language." It is rare today for new applications to be written in COBOL, and the dwindling number of existing COBOL/CODASYL applications is being steadily replaced by versions written in contemporary languages such as C, C++, or Java, or scripting languages such as PHP or Perl, using SQL (or SQL-based programming tools) for database access. This is due to the much more productive database programming capabilities and associated benefits that these combinations of technologies offer. In fact, COBOL and CODASYL are no longer taught in college and university computer science departments, and the number of programmers who are conversant with COBOL and CODASYL programming has become quite small and is shrinking rapidly. Thus, in addition to the SSA's being dependent on a set of low-productivity technologies, it is likely to become increasingly difficult for the agency to recruit and retain the pool of IT developers needed to meet its current and future IT needs.

In summary, these issues represent major challenges for the SSA going forward. Specifically, because such outdated technology is used for the

SSA's master database, it raises barriers in terms of the feasibility of linking it to new Internet applications. In addition, it causes applications to be unavailable for 24/7 access, because this technology requires downtime for updates in batch windows (see below). Some software for MADAM is written in assembly language and so closely coupled to the operating system (OS) that any changes in the OS require testing of the database. Finally, the cost and increasing scarcity of relevant custom programming expertise is making the maintenance of this technology increasing difficult and problematic. The SSA is not unaware of these issues, which have also been raised in earlier reports and studies (some are cited below). Indeed, the SSA reported to the committee that it was exploring converting to a newer technology, but the committee nevertheless has concerns about the conversion process, as discussed below.

MADAM and the MADAM Conversion Process

The SSA's Master Data Access Method, MADAM, is a database system that was developed in-house by SSA staff in the early 1980s when the SSA converted from tape to disk storage for its data sets. In 1986, the congressional Office of Technology Assessment (OTA) issued a report entitled *The Social Security Administration and Information Technology*. As part of that report, OTA expressed concerns about the use of MADAM instead of a commercial database system:

> The most controversial accomplishment of the data integration effort is perhaps the Master Data Access Method, or MADAM, the file management system that SSA developed when it converted from tape to disk storage. Many experts thought that SSA should have sought or adopted off-the-shelf software for this purpose which would be maintained by vendors, rather than developing its own which it must maintain (that is, improve, modify, and update). MADAM may well be incompatible with future mainframe operating systems, database management systems, and fourth generation languages. The SSA incurs future risks of incompatibility and long-term maintenance costs. In the short term, there are also risks and costs. MADAM is apparently a very complicated and poorly documented system, so that only a small group of people are sufficiently knowledgeable to operate it, yet it is the basis of the SSA's data management. This constitutes a particular vulnerability to smooth operations if there is any short-term emergency, sudden work force reduction, or drastic reorganization.[7]

[7]U.S. Congress, Office of Technology Assessment, *The Social Security Administration and Information Technology*, OTA-CIT-311, Washington, D.C.: U.S. Government Printing Office, 1986, p. 43 (NTIS Order PB87-136834), also available at http://www.ssa.gov/history/pdf/ota86.pdf, accessed June 20, 2007.

Despite those concerns, the committee notes that MADAM is still functioning, 20 years later. However, the vulnerability factors, including workforce issues that OTA noted, still pertain.

MADAM is layered on top of IBM's Virtual Storage Access Method product (VSAM is IBM's indexed file management package, based on the use of B-trees to support searching[8]), which allows it to run in a Parallel Sysplex environment. It appears, however, that updates to the central MADAM database are done primarily (if not exclusively) in a batch mode. New data (for example, batches of earnings reports) are first inserted into CA-IDMS and then batch loaded into MADAM during a nightly batch window (based on information that the committee received from the SSA, this window is 1:00 a.m. to 5:00 a.m. during the week and longer on the weekends). The large batch window is required to update the MADAM database daily; this scheduled downtime makes 24/7 operation of online services impossible without the implementation of some sort of complex caching and merging mechanism that could allow online services to overlap the batch window.

The continued use of the CA-IDMS/MADAM combination of technologies represents a significant barrier to the SSA goal of providing improved and expanded electronic services to users. One issue is that Java Platform, Enterprise Edition (J2EE) application servers, such as the IBM WebSphere product that the SSA is using to deliver new Internet applications, typically communicate with databases using Java and SQL, not through COBOL and CODASYL record-level operations. Connecting WebSphere applications to CA-IDMS and MADAM will involve custom adaptors that are difficult to program against from within a J2EE environment and that cannot reap the usual benefits of the productivity tools and technologies available to "normal" J2EE developers.

It is also worth noting that, aside from the software issues, the continued use of CA-IDMS and MADAM serve to lock the SSA into a particular expensive hardware and software combination. Both pieces of software only run on high-end IBM mainframes, making it impossible for the SSA to take advantage of potentially lower cost alternatives such as scalable Unix clusters or parallel database systems such as IBM DB2/PE, Oracle 10g, or NCR Teradata, among others.

One justification sometimes offered for the continued use of MADAM is its size. The argument is that the SSA databases are so large that they can only be handled by custom software. Although this argument may have been valid 20 years ago, the SSA databases are no longer uniquely

[8]B-trees are useful for exact match and range retrievals, such as looking up someone's Social Security account using the SSN as the key for the B-tree. See R. Ramakrishnan and J. Gehrke, *Database Management Systems*, Ch. 9.

large by modern enterprise standards. As of 2005, the SSA's programmatic databases had a total size of about 30 TB[9] (a terabyte is 1 trillion bytes). Proprietary concerns make it difficult to obtain published data on commercial enterprise database sizes and workloads in order to identify those that rival the SSA's database in terms of size and number of transactions. However, published data indicate that commercial systems handle very large relational databases, and relational databases with very large workloads. By way of comparative example, the customer database for Verizon, a major U.S. telecommunications company, consists of 50 billion records and 47 TB, and this database does not include data concerning the company's wireless customers.[10] Verizon's database is currently housed on a cluster of Windows Servers running Microsoft SQL Server—using commercial off-the-shelf relational database technology. Moreover, the annual Winter survey[11] lists a number of commercial relational database instances in the 20 TB to 100 TB range, and these databases are all managed by standard relational database products, each typically housed on Unix or Windows clusters. In terms of system load, SSA's workload approaches 50 million CICS transactions a day.[12] The 2005 Winter survey indicates that standard relational database systems in use at telecommunications companies and banks typically handle well over 1 million transactions per hour, with one system in the report having a peak workload of 28 million transactions per hour. This system handles this transaction load using DB2 running on Unix. It is the committee's opinion that the capabilities of commercial relational database products in 2007 have now overcome any technical rationale that might justify the SSA's continued reliance on decades-old, custom data-management technology.

Selecting a MADAM Conversion Strategy

The SSA has taken some steps to explore MADAM alternatives. A 2002 report describing the SSA's data architecture plans states that "efforts will be made to determine what portions of the [MADAM] data could be

[9]See Social Security Administration, *Information Resources Management Strategic Plan (2005)*, p. 168, available at http://www.socialsecurity.gov/irm/IRM_2005.pdf, accessed June 20, 2007. The SSA has a total of 110 terabytes of mainframe data stores, and the SSA's client-server data stores maintain 80 terabytes (ibid., p. 5).

[10]Personal communication from Michael Brodie, Distinguished Architect, Verizon Communications, to David J. DeWitt and Michael Carey, e-mail, Feb. 20, 2006.

[11]This survey is available from Winter Corporation, Waltham, Mass., http://www.wintercorp.com.

[12]The SSA reported that, in 2004, its mainframe and client-server databases had supported some 42 million individual data transactions a day. Social Security Administration, *Information Resources Management Strategic Plan (2005)*, p. 5.

managed by DB2."[13] The same language appears on page 130 of the 2005 SSA *Information Resources Management Strategic Plan.*[14] The scale and scope of the committee's investigation did not allow for an in-depth analysis of what progress has been made. However, it is clear that, given the state of relational database technology in 2007, the actual answer today to the question of what portions of MADAM data could be managed by DB2 (or another modern database solution) is simply, "All of it."

As part of this study, the committee requested documentation for MADAM and any available documents related to the SSA's ongoing MADAM-to-DB2 conversion process, especially the results of SSA pilot or feasibility studies. The following discussion is based on the documents received by the committee and conversations with SSA staff.

During the process of selecting its conversion strategy for MADAM, the SSA appears to have relied on one contractor for technical expertise. In January 2003, the SSA took delivery of a study entitled *MADAM Alternative Evaluation*, conducted by the consulting firm YL&A.[15] This study focused on alternative strategies for converting the SSA's master files, which are currently managed by the MADAM software, to DB2, IBM's relational database system product. After reviewing the MADAM database system and related applications software in some detail, the YL&A consultants concluded that:

> The MADAM, AIF, and Builder/Spreader software is a collection of custom code written in IBM assembler language (ALC). It is tightly coupled with the operating system, is extremely complicated, and can only be supported by a limited number of extremely talented application developers. In fact, MADAM is so tied to the operating system that any change to the operating system has to be tested for its impact on MADAM. There is some fear that future operating system changes may render MADAM unusable and the talent need [sic] to remedy the situation may no longer be accessible.[16]

The *MADAM Alternative Evaluation* report begins with a brief overview of the different master files and the advantages that switching to a commercial relational database system such as DB2 would provide, even if DB2 proves to be slower for certain operations. Section 5 of that report proposes five alternative proof-of-concept (POC) studies that the SSA

---

[13]Social Security Administration, "Future Database Management Sub-Architecture," *SSA Data Architecture*, Version 1.0, December 2002, p. 30.

[14]Social Security Administration, *Information Resources Management Strategic Plan (2005)*, p. 130.

[15]YL&A, *Social Security Administration—MADAM Alternative Evaluation*, Springfield, Ill.: Jan. 22, 2003 (hereafter cited as YL&A, *MADAM Alternative Evaluation*).

[16]YL&A, *MADAM Alternative Evaluation*, p. 9.

could undertake to evaluate the difficulty of replacing MADAM with DB2 and recommends selecting the Supplemental Security Income Records (SSR) database for a POC conversion effort.

The same report also considers alternative approaches to converting the SSA's various applications (for example, the programs that do things like issuing monthly checks). It appears that the bulk of these applications are currently written in IBM assembly language[17] and make use of an abstraction layer known as the Application Interface Facility (AIF) that is built on top of the various MADAM files.

The existence of such an interface provides a number of alternative ways of converting the actual applications. For example, one could conceivably rewrite the AIF software to make DB2 calls instead of MADAM calls. This would make it theoretically possible to run the existing applications without modification. Although this sounds very appealing at first glance, it in effect turns DB2 into a MADAM simulator. This is not a useful approach in the long term because it would forgo many of the reasons for doing the conversion in the first place, and it almost certainly would sacrifice long-term flexibility and performance for short-term gains. (See the next section for more on architecture.)

A second, 2004 report from YL&A entitled *SSR Master File to DB2 Conversion Deliverables Detail Document*[18] describes the results of the POC study recommended by YL&A in its January 2003 report to convert a portion of the records in the SSR master file to DB2. Unfortunately, this second report does not contain key details of the POC study (for example, the DB2 schema used); hence it is not possible for the committee to evaluate the results of the study or the approach adopted. Certain things suggested in the 2004 report do, however, cause concern, as elaborated below.

As noted above, the POC that YL&A and the SSA executed adopted the philosophy that any MADAM-to-DB2 conversion should have little or no impact on existing applications. As a consequence of this decision, the resulting database design was severely compromised with *individual* SSR records "stored in as many as 46 DB2 tables."[19] Such an approach is almost certainly going to experience poor performance, suffer from numerous potential update anomalies, and almost inevitably be the source of numerous long-term complications if actually adopted. The proposal of such a design in the first place is very surprising and illustrates the inadvisability of adopting a MADAM-to-DB2 conversion approach that does

[17]YL&A, *MADAM Alternative Evaluation*, p. 13.

[18]YL&A, *SSR Master File to DB2 Conversion Deliverables Detail Document*, Springfield, Ill., July 1, 2004.

[19]YL&A, *SSR Master File to DB2 Conversion Deliverables Detail Document*, Springfield, Ill., July 1, 2004, p. 14.

not include a rewrite of the actual applications to make DB2 calls directly. (Furthermore, if 46 tables were really needed, the use of views might be preferable, instead of actually storing an individual record 46 times.)

Another issue is the apparent failure of the YL&A studies to consider other large-scale relational database management system (RDBMS) technologies. There are several other vendors in addition to IBM (DB2), and these vary in their performance, functionality, and scaling capabilities. However, the YL&A report does not enumerate the SSA's requirements and evaluate other leading vendors' systems[20] against them, thus not enabling the SSA to make the most informed selection.

Migrating legacy databases of the size and vintage of MADAM is fraught with challenges.[21] While long-established enterprises (such as major telecommunications companies) have faced the challenges of migrating legacy applications and databases, seldom if ever are databases migrated from one platform to another unless the target database is a close match to the source database, in which case, the process is relatively straightforward. Consulting firms and DBMS vendors (for example, IBM, Oracle, and Microsoft) are in the business of database migration with database-migration tools and factories. While these factories and tools address the database migration, they are less able to deal with the greatest cost and complexity, namely, all software above the database layer—middleware, application programming interfaces (APIs), and thousands of calls throughout known and unknown applications that access the database. Hence, database migration for similarly structured targets and sources is relatively straightforward, while migrating the software stack above it is almost never straightforward. It is unlikely that the structure of SSA's MADAM is close to what would be required of a new database to support SSA's e-government and related requirements. Hence, it is unlikely that any conventional database-migration strategies or tools would be applicable.

With very rare exceptions, major companies have not attempted direct translation of mainframe databases from one platform to another without significantly affecting applications. The rare cases involve large VSAM files using powerful tool sets for VSAM to significantly manipulate the

---

[20]For 2005, the leading RDBMSs, based on worldwide market share, were as follows: Oracle (44.6%), IBM (21.4%), Microsoft (16.8%), Sybase (2.5%), and NCR Teradata (2.9%). The rank order was the same in 2004. "IDC Reveals Steady Growth for Worldwide RDBMS Market as Top 5 Vendors Hold Their Positions," Press Release, Framingham, Mass.: IDC, May 24, 2006.

[21]See M. Brodie and M. Stonebraker, *Legacy Information Systems Migration: The Incremental Strategy*, San Francisco, Calif.: Morgan Kaufmann Publishers, 1995, for an extended discussion of issues to consider when contemplating the challenges and risks of a large-scale data migration of the sort that the MADAM-to-modern-DBMS migration poses.

files prior to their being imported into Information Management System (IMS) or DB2 databases aided by migration tools designed by IBM and others for the migration of VSAM to IMS or DB2. Most successful cases have involved completely replacing the legacy applications and databases with entirely new applications and databases. In such cases, legacy databases are not migrated in toto into new databases, but instead the necessary information content is extracted and incorporated into the new databases. The legacy applications and databases are then capped on a specific cutoff date, after which time all new transactions are redirected to the new applications and databases. For operational, regulatory, and other reasons, the legacy databases and their applications are usually maintained in case access to information that had been in use prior to the cutoff date is ever needed. The new applications and databases then become the systems of record for all transactions from the legacy system cutoff date or the new systems production date forward. This methodology eliminates the challenges involved in attempting to migrate the totality of all legacy data, structures, and other artifacts from legacy databases to new databases. It also frees the new databases and applications to focus on current and future requirements, similar to those associated with the SSA's e-government strategy.

Finding: In 1986, the Office of Technology Assessment alerted the SSA to the technology and technological risks inherent in its Master Data Access Method, or MADAM. Today, MADAM and the SSA's current data-management approach continue to pose increasing risks. The approach faces increasing limits in the availability of staff who understand and can support the SSA's technologically obsolete, custom solution. In addition, the approach precludes the use of valuable new technological capabilities and requires interruptions in service for batch updates, both of which impede the provision of desirable new e-services.

Recommendation: As it makes decisions about future directions for its database technology, the SSA should give considerable weight to the implications of those decisions for the effectiveness and efficiency of current and future electronic service delivery and should be open to the introduction of new technologies.

Architecture

In addition to specific database challenges, architectural considerations are a key component of a successful electronic services strategy. It is beyond the scope of this report to recommend a specific architectural approach. In principle, however, the SSA will need to develop a data-management strategy and architecture. Doing so entails developing a data-management architecture that includes appropriate storage to meet the requirements of the organization's e-government strategy and other operations. In addition, it entails the development of a migration strategy and supporting architecture for capturing the needed pre-replacement information and making it available as part of the new data-management architecture. Where possible, the replacement should be incremental in order to minimize risk and to minimize interruption of the agency's workload and operations. The data-management architecture should be an integral part of the next-generation architecture.

While the SSA has some modern technology components—for example, its Web site and security practices—the core technologies, such as data management, are outdated and pose significant risks and limitations to the SSA's operations and impose constraints on the SSA's ability to achieve an electronic services strategy comparable with those in the private sector (see Chapter 2). In addition to data management, user interaction, and of course security, architectural considerations loom large in the necessary transition. The committee does not recommend a particular future architecture, but a number of questions should be addressed as the SSA moves forward. What would that target architecture be? How would the SSA migrate from the current architecture to the target incrementally without disrupting its operations and at the same time improving productivity, responsiveness, and new services and lowering cost? The data-management issue should not be addressed without a target architecture. The data-management challenge is an integral part of a larger challenge, the architecture.

The committee acknowledges that it is easy, from the outside, to counsel a radical reengineering of an outdated legacy system, while it is very much harder to build a case for the major expense and downside risks entailed in the retirement of such a system with its large historical investment. The SSA, in addition to the technical considerations outlined here, will also need to take into account congressional attention and the likely inevitable congressional hearings and newspaper headlines that will ensue if the reengineering efforts result in even a single month in which Social Security checks do not go out on time. Because the committee did not receive detailed information about the current nature and status of the SSA's MADAM conversion effort, it did not reach conclusions as to whether any ongoing activities should stop. However, as described above, the committee does have concerns about some of the technical advice that the SSA has received regarding conversion. The SSA thus should look broadly for advice as it plans and executes its MADAM conversion effort and architectural overhaul.

In seeking broader technical advice, the SSA might, for example, establish an external advisory board consisting of experts from the database software industry, several large commercial enterprises (financial services and telecommunications companies) that have deployed large relational databases, and academics to oversee the MADAM conversion effort. Relying on a comparatively narrow range of expertise to architect such a massive conversion effort creates needless risks for successful outcomes. In many ways, the SSA is at a critical crossroads. The way forward is difficult and risky, but standing still is risky as well. The committee believes that the best way to move forward is to ensure that the SSA has the best possible technical advice—both to guide the agency and to provide context and insights on managing the inevitable glitches that will occur.

Finding: The scope and scale of the challenges that the SSA faces with regard to its database conversion strategy merit the input of a broad range of expertise.

Recommendation: In continuing to develop its conversion strategy and long-term services strategy, the SSA should draw on a broad range of technical expertise—including but not limited to database software experts, software engineers, software security experts, financial services experts, large-scale commercial service providers, and systems architecture experts—and put systematic mechanisms in place so that it can hear and learn from outside advisers.

Other Applications

As indicated above, the key programmatic SSA applications[22] are the following:

- *Title II:* Administering disability, old age, and survivor benefits;
- *Title XVI:* Administering Supplemental Security Income;
- *Earnings:* Recording of annual wage reports;
- *Enumeration:* Allocation and verification of Social Security numbers; and
- *Disability:* Determination, control, and tracking of disability claims.

---

[22]Factual information in this subsection is from material presented to the study committee by SSA and Disability Determination Services speakers during study briefings and from other material provided to the committee.

Of these key applications, all but Disability are implemented as batch applications against the CA-IDMS and MADAM databases, with CICS being used to drive "green-screen" application interfaces.[23]

Over the past few years, the SSA, as part of its Electronic Disability process (eDIB) and paperless claims effort,[24] working in partnership with state Disability Determination Services (DDS) offices, has implemented an electronic version of the disability determination process using WebSphere and DB2. With this new system, users can initiate claims either over the Internet through the SSA Web site or at their local disability office. As a claim is assembled, all the supporting medical evidence (for example, x-rays, doctor's and laboratory reports, and so on) is digitized and stored in the DB2 database (using DB2 large object capabilities) along with the other details of the claim. At the time of this writing, the eDIB application was still in the process of being rolled out state by state. Use of eDIB will reportedly cost $850 million over a 7-year life cycle, and the SSA expects to reap $1.3 billion in total savings (including postage and file-storage costs) from eDIB and to cut more than 3 months from the average time required to process a disability claim.[25]

Although the eDIB effort is certainly a step in the right direction, the implementation is basically an electronic "copy" of the formerly paper-based process. The electronic folder that is ultimately assembled using the application process closely resembles the paper folder that it replaces. This is a good first step, and the approach provides a natural online transition for users of the system, but it misses a number of opportunities for enhancement. For example, with the current electronic folder, all supporting documents—including not just multimedia artifacts such as x-rays and photographs, but also all doctors' reports, laboratory reports, and so on—are converted to images encoded in the Tagged Image File Format (TIFF) format. This approach does succeed in archiving the documents but makes it impossible to index and search them without resorting to

---

[23] "Green screen" is common usage to refer to the monochrome-display cathode ray tubes (CRTs) used as computer monitors, predominantly in the 1970s and 1980s. A green phosphor was used, producing green characters on a dark background; some early CRT monitors used orange phosphors.

[24] The eDIB process is intended to provide better and more timely service to claimants, in part by replacing voluminous paper case files with electronic files. In contrast to transferring the electronic case files used for "paperless" claims processing, locating and physically transferring paper files when a disability claimant requests a hearing used to take a month or more. The SSA describes its eDIB and paperless processing approaches in Social Security Administration, *Information Resources Management Strategic Plan (2005)*, available at http://www.socialsecurity.gov/irm/IRM_2005.pdf, accessed June 20, 2007.

[25] Mary Mosquera, "Case Files Travel Lighter, Faster," *Government Computer News*, Oct. 9, 2006, available at http://www.gcn.com/print/25_30/42177-1.html, accessed June 14, 2007.

optical character recognition of the image files, a process that will introduce errors. In contrast, if doctors' reports were captured and stored in the database in a format such as Word documents or Portable Document Format (PDF), they could readily be indexed and searched. This would greatly enhance the accessibility and usability of the data in the eDIB database. In the future, as machine-readable standards for electronic patient records take the place of handwritten notes and other unstructured documents, it would be useful to be able to store such records directly in the disability database.[26]

Web and Internet Services

The SSA's public Web site (http://www.ssa.gov) is a primary access point for many users of electronic services. Although the previous components of the agency's IT infrastructure all play a role in electronic service provision—the database technologies and architectures in particular are key—the Web site typically provides the first impression and the bulk of the interaction experience for most users. Thus, attention to interface considerations, information architecture, and an intuitive and informative user experience is critical.

Although the SSA was one of the first government agencies to have a Web site, it is now several years behind what commercial e-commerce and financial services sites are providing today (see Chapter 2 for more on expectations and offerings of world-class financial institutions). The current SSA Web site is largely a site for dispensing information about SSA programs, with capability for some (partial) online benefits applications and some retirement-planning calculations that require user-provided earnings and estimated benefit data. Thus, for example, a very basic operation that one might expect to find, being able to view one's earnings history, is not supported at the time of this writing. The problem is more a policy issue than a technical one, and it may stem in part from a 1997 episode, when the SSA made an initial attempt to provide online services without completely thinking through the approach to security (specifically, only the SSN, mother's maiden name, and state of birth were used for account authentication) and was strongly criticized as a result. However, given the widespread public adoption of online and secure access to bank and brokerage accounts that has occurred during the intervening

[26]For more on general challenges relating to storing or archiving electronic records, see the Computer Science and Telecommunications Board's recent report for the National Archives and Records Administration: National Research Council, *Building an Electronic Records Archive at the National Archives and Records Administration: Recommendations for a Long-Term Strategy*, Washington, D.C.: The National Academies Press, 2005.

period of about 10 years, it seems possible that there may have been a change in the public's attitudes about trade-offs between privacy protection and the benefits associated with having online access to confidential financial information. Accordingly, the SSA should continue to monitor public attitudes toward online access to sensitive personal information in order to determine if and how the SSA might furnish individual records and data online in ways that are consistent with contemporary practices, that present minimal and acceptable risk, and that are acceptable to the users of its services.

The current SSA Web site is limited in several additional ways. First, only a subset of the information on the SSA site is available in a language other than English. Furthermore, the online application for Social Security retirement and disability benefits is only available in English and, at the time of writing of this report, was only accessible using particular browsers.[27] The benefits application pages explicitly state that there is no support for users of Apple or WebTV systems. As noted in Chapter 2, this situation should not be allowed to persist, as this level of services falls short of the current state of cross-browser and cross-platform compatibility found in such commercial sectors as financial services and telecommunications.

Future Directions for Software Development

A number of trends in the computer software-development and applications arenas will be relevant to the SSA as it updates its systems and develops its strategy for electronic services. These trends, which should be watched and leveraged over time,[28] include the following:

- The Service-Oriented Architecture (SOA)[29] wave,

---

[27]On October 24, 2006, the SSA Web site section "Social Security Claims" at https://s044a90.ssa.gov/apps6z/ISBA/main.html had the following message: "Note to Mac and Web TV Users: This Internet service does not support Mac or Web TV browsers. If you are a Mac/Web TV user, please call one of the toll-free numbers listed on the page. Our representatives can arrange for your application to be taken over the phone or in person at a Social Security office."

[28]Names of specific companies and technologies are offered simply as examples. The committee does not necessarily endorse or recommend any particular company or technology.

[29]See, for example, M. Huhns and M. Singh, "Service-Oriented Computing: Key Concepts and Principles," *IEEE Internet Computing* 1(9):75-81, 2005; G. Alonso, F. Casati, H. Kuno, and V. Machiraju, *Web Services: Concepts, Architectures, and Applications*, Berlin/Heidelberg: Springer-Verlag, 2004; M. Singh and M. Huhns, *Service-Oriented Computing: Semantics, Processes, Agents*, West Sussex, England: Wiley, 2005. See also the work of the OASIS SOA Reference Model technical committee at http://www.oasis-open.org/committees/tc_home.php?wg_abbrev=soa-rm, accessed June 20, 2007.

- Software as services and resulting value-added service and third-party software opportunities,
- The proliferation of Extensible Markup Language (XML) and its impact both on search and on record format standardization efforts,
- Advances in data mining and machine learning techniques,
- Continued advances in commercial database technology, and
- Advances in interactive applications.

SOA is one of the most highly touted trends in enterprise IT software today. Like the SSA, virtually every large enterprise today runs on a heterogeneous collection of software applications, some off the shelf (for example, SAP, PeopleSoft, Siebel, Oracle) and some in-house (and often mainframe-based) applications that would be difficult or expensive to replace. In order to automate more of their business processes to offer new and enhanced services to various stakeholder groups, enterprises often need to develop new applications that integrate functionality and data from multiple existing applications. Responding to this need, software research in the 1990s focused on "megaprogramming,"[30] workflow systems technology, and distributed component-based programming models. The fruits of that research—and of the corresponding industrial activity in the area—include XML-based Web services and new, high-level programming approaches and tools for building new applications by composing existing services (also known as "programming by composition"). The SOA trend and its associated tools are likely to be very relevant to the SSA, providing high-productivity tools for service-enabling the agency's existing applications and for using and/or reusing the resulting component services in constructing new online services and internal applications.

A companion trend to the one involving SOA and high-productivity tools for service composition is the availability of tools for business activity monitoring (BAM) and/or business process analysis (BPA). These tools support the modeling, monitoring, and gathering of analytic information about the day-to-day operations of business processes. Such tools, which are rapidly maturing, could enable the SSA to gain visibility into the efficiency and performance of its business processes, providing information that could then be used to diagnose bottlenecks and other problems and to tune or adapt its processes over time to provide better service to its users and clients. Disability-claim processing, a complex and lengthy SSA process, seems to be a particularly appropriate target for such monitoring and analysis tools. Other tools will be able to analyze process

[30]G. Wiederhold, P. Wegner, and S. Ceri, "Towards Mega-Programming," *Communications of the ACM* 11(35):89-99, 1992.

definitions in order to identify process defects that could lead to incorrect outcomes.

Another related and very relevant trend is the "software as a service" trend. Rather than building and selling software packages that enterprises must then install, manage, and maintain or upgrade, a number of companies are starting to offer "software as a service"—they build and manage vertical applications that are available to enterprises through a combination of Web services and interactive Web pages.[31] Another relevant example is in the tax preparation software market, where companies including H&R Block and TurboTax are now offering tax preparation online as well as through client-installed software packages. The types of services and interactive client applications that these industries are now offering are not unlike the services needed by SSA partners and clients, particularly for complex functions such as disability-claim preparation and planning, "what if" planning for various benefits (such as online retirement planning, assistance with prescription drug program selection, and impact analysis of change in employment status), representative payment, annual wage reporting, and so on.

Trends in data formats and standardization thereof, notably centered on XML, are very relevant to the SSA. In the past several years, XML has taken the machine-to-machine Internet data-transfer field by storm, not unlike the impact that HyperText Markup Language (HTML) had on the machine-to-person side of the Internet. XML is now central to data interchange both within and between enterprises. As a result, it is now also at the heart of many efforts for standardizing electronic documents for use in a wide variety of fields, including insurance, law, manufacturing, business, and of course health care (e.g., the Health Level Seven, or HL7 initiative[32]). Continued adoption of XML and associated schema standards for data interchange appears likely to offer the SSA significant future opportunities to streamline and automate its interactions with many of its user communities, such as other government agencies and nongovernment business partners.

Another significant data-format trend is the use of XML as a file format for office automation software, such as the Organization for the Advancement of Structured Information Standards (OASIS) Open Document Format for Office Applications and the new Microsoft Office open

---

[31]SalesForce.com is a leading example of this class of software vendor—it provides a leading customer relationship management (CRM) application via this "software as a service" model. The significant traction that SalesForce.com has in the CRM marketplace is an indication of the potential for this model.

[32]HL7 is one of several American National Standards Institute (ANSI)-accredited Standards Developing Organizations operating in the health care arena. HL7's domain is clinical and administrative data.

XML formats. These trends could open up many types of electronic documents—an SSA example might be various reports related to disability claims—to broad and efficient searching (subject, of course, to stringent access control in order to protect privacy) and more advanced XML-based semistructured content search technologies as they are developed. Such technologies should provide the SSA with powerful search capabilities that can be leveraged for disability-claims processing and related services in the future.

Moving to a different technology area, the past decade has seen great advances in the area sometimes referred to as Knowledge and Data Discovery. Significant advances have been made in the areas of data mining and machine learning, giving rise to a number of new approaches to mining very large sets of data, building and maintaining classification networks, and so on. These advances are being exploited today in a number of industries and government agencies that face problems not unlike those that confront the SSA. The insurance industry is successfully using these advances to speed claim analysis and provide enhanced automatic fraud detection, while the financial industry is using these advances to automate much of the loan origination process (such as for mortgage lending). As the SSA seeks to improve the quality and timeliness of service to its constituents, it stands to benefit from the use of these advances as well. As an example, it seems well within the capabilities of this technology for the SSA to automate fully the classification of "simple" disability claims. Presumably the SSA would wish to ensure some level of human monitoring of such automated classification, but efficiencies would certainly still be obtained. This technology could also be put to work on the other end of the process to look for patterns indicating potential fraud.

Last but not least, advances continue in the database-management field as well, and the SSA is definitely a strong candidate for benefiting from these advances. As discussed elsewhere in the report, relational database system technology has made huge strides in the 35 years since its inception, having gone from theoretical speculation in 1970 to serious commercial offerings in 1980 to a booming industry in 1990 to the heart of almost every serious enterprise's IT infrastructure today. Telecommunications and financial services companies manage many terabytes of data today using commercial off-the-shelf relational database management systems. Relational database technology today is on the verge of commoditization on the lower end, and on the higher end there are a number of high-performance, highly scalable systems available that provide integrated replication services for failover and disaster recovery. Moreover, most of these systems provide support for advanced "large" or "multimedia" data types, including text, images, geospatial data, and XML—essentially every imaginable data type that the SSA needs to store, manage, and search. Also as mentioned earlier, database technology limitations are definitely no longer a reason for the SSA to rely on the use of in-house data management solutions.

As with the case of best practices described in Chapter 2, trends in technology development and deployment should also be monitored carefully for their potential relevance to the SSA. There are a number of technologies on the horizon—this section mentions several exemplars—that appear to be gaining significant traction in other IT sectors and that may be of great use to the SSA. These should be monitored and adopted as appropriate—they may not all mature at the same rate or at the same time, or even soon.

SECURITY, PRIVACY, AND AUTHENTICATION

Like other institutions, the SSA faces an onslaught of information and computer security challenges, including attempted break-ins, theft, tampering, and phishing. And, as with any other large organization, the SSA will also inevitably face challenges in dealing with the problem of having insiders participate in facilitating violations of privacy and security, the so-called insider problem.[33] In recognition of these challenges, there is understandably a very strong security focus underlying the SSA's IT efforts. In the committee's view, the SSA clearly takes the problem of security very seriously and appears to have adopted best industry practices. The SSA views its user community as consisting of four distinct groups: employees and trusted individuals, government agencies, nongovernment business partners, and the general public. Identification, authentication, and authorization are handled differently for each of these groups.

Internally, the SSA security team is organized into four subteams, as follows:

- *The policy group* sets security policies.
- *The standards group* creates standards that embody the policies set by the policy group.
- *The execute group* implements the organization's security policies in software and architecture.
- *The oversight group* ensures that SSA IT projects use the appropri-

[33]"White Paper: Cyber-Security and the Insider Threat to Classified Information," a non-reviewed summary prepared by the National Research Council's Computer Science and Telecommunications Board, 2001, available at http://cstb.org/whitepaper_insiderthreat, accessed June 15, 2007. See also M. Bishop, *Computer Security: Art and Science*, New York: Addison-Wesley Professional, 2002, p. 21.

ate tools and that the resulting implementations match the established policies.

The SSA also does regular penetration testing of its Web site; this is one of the responsibilities of its Intrusion Protection Team, which is made up of both SSA employees and outside penetration testing specialists. In addition, the SSA has a contract with an external vendor (IBM Managed Security Services) for providing real-time intrusion-detection services.

These security approaches are aimed at, among other things, protecting the integrity and confidentiality of the data that SSA maintains. In terms of SSA's approach to privacy from a policy standpoint, the SSA Web site states that with regard to government information exchange, "The privacy of all personal information SSA maintains in its databases is protected and controlled by a number of Federal statutes, including section 1106 of the Social Security Act, the Privacy Act of 1974, section 6103 of the Internal Revenue Code, and related Social Security regulations and policies."[34]

In the case of the SSA's electronic services in particular, the issue of privacy typically revolves around how and why individual data are accessed or disclosed and to what extent individuals can control access to their data. As is often the case, privacy overlaps security in that confidentiality mechanisms provide the tools used to control access to data and to monitor to whom the SSA allows access to customer data. The SSA collects financial and medical data about individuals, as well as customer service data. The former support the SSA's goals of financial stewardship and service provision; the latter improves the SSA's ability to provide that service. There are likely cases in which those who need access to customer service data will not need access to the financial or medical data and vice versa. Thus, keeping these types of data separate and secure unless, perhaps, when there is a programmatic need for both types of data (for example, when paying medical bills) can serve to protect privacy while not compromising needed service provision.

Authentication

Authentication—both technological measures that authenticate individuals and ways to perform authentication that respect privacy—was discussed at length in a previous National Research Council report that

---

[34]See http://www.ssa.gov/gix/privacyinfo.html, accessed June 20, 2007. In addition, the SSA's privacy policy for its online offerings is available at http://www.ssa.gov/privacy.html, accessed June 20, 2007.

was partially funded by the SSA.[35] Authentication issues, including the existence and use of a unique identifier such as the SSN, continue to be highly controversial and continue to hinder progress on providing a wide range of electronic services. For example, comprehensive authentication across multiple applications would allow the SSA's clients to authenticate once per session, without having to reenter their data in order to conduct transactions related to separate applications or programs. The committee conjectures that part of the problem in moving forward with comprehensive online authentication solutions is the recognition that any security or privacy breach may result in chastisement by Congress. Furthermore, commercial sites generally factor the financial cost of a certain potential level of fraud into their basic cost structure. Commercial organizations reason that it can be less expensive to tolerate certain improbable potential losses than to try to prevent them at any cost. As with other financial institutions that face trade-offs between security and accessibility, in most cases the costs of security breaches to the SSA may be more a matter of reputation than of finance. Although the risk-cost or risk-benefit analyses can be challenging, it is nevertheless important for the SSA to weigh the agency and public benefits from expanding electronic service functionalities against the incremental risks and costs.

For its current online suite of services, the SSA's solution to the authentication problem is to do initial password assignments through the U.S. mail. (And, indeed, some financial institutions still use a Postal Service channel for initial verification when setting up online authentication for an individual.) In June 2006, individuals *already receiving benefits* could apply online for a temporary password request code (PRC) by providing their date of birth and SSN through a secure transmission. (The SSN itself cannot be used as a secure password/secret; see Box 3.2 for more on the SSN as an identifier.) At that time, the SSA Web site indicated that a person could expect to wait at least 15 days for a temporary PRC to be mailed to his or her address of record. Once a beneficiary has a temporary PRC, he or she can use it to create a permanent password. Logging in allows a person to check personal information and benefits; see his or her address, telephone number, direct deposit, and Medicare and payment information; change the address and/or telephone number in the SSA's records; request or change direct deposit; start direct deposit of checks or change

[35]National Research Council, *Who Goes There?: Authentication Through the Lens of Privacy*, Washington, D.C.: The National Academies Press, 2003.

BOX 3.2
The SSN as Identifier and Authenticator

Social Security numbers (SSNs) are nine-digit numbers whose first three digits are related to the geographic location where the SSN was issued. The SSN concept was originally created in 1935-1936 to keep track of workers' earnings. After Executive Order 9397, issued in 1943, required federal agencies to use the SSN to identify persons in any new federal systems-of-records systems, its use as an identifier in these systems grew, but slowly, for nearly 20 years. Then, in 1961, the Civil Service Commission began using the SSN to identify federal employees; the Internal Revenue Service began using it as the official taxpayer identification number the next year. As computers came into more widespread use, the use of the SSN as an identifier by federal, state, and local governments and private-sector organizations grew very rapidly.

No legislation prohibited broad use of the SSN. Indeed, in 1970, congressional concerns about welfare fraud and unauthorized workers led to amendments to the Social Security Act authorizing the Social Security Administration (SSA) to assign SSNs to legally admitted noncitizens entering the United States and to people applying for or receiving federal benefits. Subsequent legislation continued to expand the use of the SSN—for example, as a condition of eligibility for federal assistance or loan programs; for use by states as part of their own tax processing, public assistance, driver's license, or motor vehicle registration functions; for use in child support enforcement; and for military Selective Service (draft) purposes. According to the SSA, there are now 27 authorized uses of the SSN as the identifier for record-keeping or matching criteria.[1]

Although numerous laws authorize and/or require the use of the SSN as an identifier, no federal law regulates its overall use. Some federal laws restrict the use of SSNs in certain programs to specific uses and prohibit unauthorized disclosure, but there is no federal law regulating or restricting the use of SSNs by the private sector.[2] Their use as a de facto near-universal identifier has proven to be problematic—resulting in easier paths to identity theft and credit fraud.

---

[1]The information in this history is from Social Security Administration, *Report to Congress on Options for Enhancing the Social Security Card* (undated), available at http://www.ssa.gov/history/reports/ssnreport.html, accessed June 20, 2007.

[2]See U.S. General Accounting Office, *Social Security: Government and Commercial Use of the Social Security Number Is Widespread*, GAO-HEHS-99-28, Washington, D.C., February 1999.

the current direct deposit to another account or financial institution; and change the password.[36]

---

[36]Information obtained from the SSA's Web site at https://s3abaca.ssa.gov/pro/passregi/passserv.shtml, accessed June 23, 2006. In June 2007, the SSA's updated Web site has new information indicating that a current beneficiary or a person who has recently applied for benefits can request a password. See https://s044a90.ssa.gov/apps6z/ACU_LDPG/

When possible, most institutions are now moving away from use of the SSN (for example, as student identification [IDs], subscriber IDs, customer IDs, and so on). Given the widespread use of the SSN and its increasing exposure to potential misuse (for example, from lost or compromised government and corporate data containing millions of individuals' SSNs and other identifying information), knowledge of an SSN should not be the only information required for an individual to authenticate himself or herself to the SSA or any other institution.

Indeed, as the SSN contains only 9 digits (making for a maximum of 1 billion different SSNs), and the number of issued SSNs exceeds 400 million,[3] there is at least a 40 percent probability that a randomly generated 9-digit number will be someone's SSN. Thus, the security risk from relying solely on the SSN for authentication is large and growing. In comparison, American Express card account numbers are 14 digits long, Visa/MasterCard account numbers are 16 digits long, and both have a 3- or 4-digit security code marked on the card to prevent fraud.[4] Additional information related to an account (for example, the billing zip code, a personal identification number, or details about a recent transaction) is often required when interacting with a financial institution in a way that could compromise an account.

The security of the SSN, given its ubiquity and comparative simplicity, is a challenge. Addressing this challenge involves not just the SSA but also the multiplicity of federal, state, local, and private-sector entities that use the Social Security number to identify individuals or accounts.

---

[3]By 2006, more than 420 million SSNs had been issued. See http://www.ssa.gov/history/hfaq.html, accessed June 20, 2007. The structure of the numbers includes a geographical area number (the first three digits) and a group number (the next two digits), then four consecutive digits from 0001 to 9999. Because group numbers are not issued sequentially for a given area, at a given point in time some nine-digit numbers are not possible as valid SSNs. The SSA publishes SSN number ranges issued to date. See http://ssa.gov/history/ssn/geocard.html, accessed June 20, 2007.

[4]The security code is not embossed on the card; therefore, it does not show up on the credit card slip. This prevents someone from obtaining the code from a receipt.

Note that a user's password is not sufficient to view his or her earnings history online. Indeed, only current beneficiaries can get passwords.

---

landingpage, accessed June 20, 2007. In 2007, as in 2006, the password request functions are not available 24/7, and the SSA's Web site indicates that the waiting period for receiving a temporary PRC by mail can be up to 15 days. See https://s044a90.ssa.gov/apps6z/ACU_LDPG/landingpage, accessed June 20, 2007.

A beneficiary's password is, however, sufficient to change the bank account to which direct deposits are made. The committee conjectures that the former may be considered not merely a security risk but also a more emotionally laden privacy risk, even as security breaches regarding the earnings history are less likely to have real financial consequences than would security breaches regarding bank accounts and direct deposits. It is unclear whether this policy is a carefully considered choice, resulting from a rigorous cost-benefit analysis, or an unintended consequence of various privacy and security policies influenced by oversight probes.

Security and Privacy in a Dynamic Environment

In addition to continuing changes in hardware, software, and the ways in which people will choose to access services, large organizations such as the SSA will also have to address how their security, privacy, and authentication policies may need to change over time. Over the next decades, the SSA will face many challenges—both internal and external—in maintaining privacy and security. The external challenges will arise from entities that wish to obtain information or access for which they are not authorized. The internal challenges will come from entities that are authorized to have access or information but the authorized entities abuse or misuse it, or even just make a mistake.

The set of policies that define security, privacy, and authentication for the SSA are controlled not only by the agency itself but also by external factors such as statutes and regulations (see Appendix D regarding some of the relevant statutes). The agency is constrained to choose mechanisms to enforce these policies in a way that complies with the goals and constraints under which it functions. In particular, the SSA's entire e-government approach must view security, privacy, and authentication policies in the context of the agency as a whole, with electronic services as part of an overall service-delivery strategy (see Chapter 4).

Numerous approaches to information system security (in this case encompassing privacy and authentication, as well) have been developed by the information security community.[37] On the basis of briefings and conversations with SSA technical staff, the committee considers that staff to be well informed in these areas and mindful of the challenges and principles involved.

The SSA consists of several components, each of which manages different sets of systems. Each part of the agency has performed security

[37]See, for example, at http://cstb.org/topic_security/ a listing of numerous reports from the Computer Science and Telecommunications Board on cybersecurity and computer security.

analyses. According to information given to the committee, the SSA has clearly weighed the risks very heavily. One caution, however, is that such an emphasis may lead to an unnecessary use of mechanisms that impair one or more security services. This can occur because security can be thought of as consisting of three types of services: confidentiality services, integrity services, and availability services. For example, the availability of information is critical for timely transactions and handling of cases. But availability carries with it the risk that the information may be disclosed to unauthorized parties. Hence, limiting availability reduces risks to confidentiality, but it increases risks to availability. The security policy must strike the correct balance between these qualities, and that balance should be determined by risk analysis and cost-benefit analysis.

The factors involved in determining approaches to security and policy include not only the SSA's internal rules, but also the laws and regulations imposed externally by Congress and by other federal agencies (such as the Office of Management and Budget [OMB]) and court rulings. Further, agreements with state agencies impose additional constraints. Thus, technical concerns are not paramount but are simply one factor to be considered along with these other factors. Clearly, none of the above is meant to suggest that the SSA should not pay appropriate attention to security risks. Both internal controls and measures to address external threats (fraud, hackers, and so on) must be considered. To be most effective, security must be built into a system from the start. Treating it as an add-on or applying Band-Aid-type measures is costly in terms of resources and risk. Although the SSA may not have taken a security-from-the-start approach in all of its previous electronic services projects,[38] this report is aimed at encouraging the agency to adopt that posture as it seeks to expand the scope of its electronic services. Admittedly the agency cannot address these challenges alone—as with other government agencies, congressional and/or executive branch (through OMB, for example), involvement could help provide a framework and broader support for SSA actions to offset the threat of potential negative reactions.

---

[38]See SSA, Assistant Inspector General for Audit, "Evaluation of the Accelerated eDIB System—Third Assessment (A-14-03-13047), December 20, 2002, available at http://www.ssa.gov/oig/ADOBEPDF?A-14-03-13047.pdf, accessed June 29, 2006. The SSA's Office of the Inspector General (OIG) found that the August 2000 *eDIB Program Management Plan* "neither addressed security nor evaluated the risks involved in eDib program development. OIG concerns were partially addressed in the November 14, 2001, *eDIb Program Management Plan*. . . . However the [2001] plan did not address the risks associated with security, fraud, hackers, and complexity of the system. Instead, the Risk Management Plan addressed development risks, which could be incurred during systems development, such as cost, schedule, integration/technical and mission. While system development risks should be considered, it is as important to address risks that relate to internal controls and security."

BOX 3.3
General Security Principles

In a dynamic and changing environment, the policy and technology questions relating to security and privacy must be revisited on a continuing basis. The following are several useful principles for any large organization developing security policies:

- *Simplicity.* The more complex a policy or system is, the more that can go wrong. Worse, complexity of interfaces leads to errors, because humans develop mental models of how systems work and base their actions on these models. If a system is too complex, the mental models either fail to capture the system behavior or take too long to develop. In either case, errors occur. It is critical to understand that this pattern applies not only to "outsiders" using interfaces to access systems but also to developers of systems, who have mental models of how the components work; to analysts, who have mental models of what to look for and how to look for weaknesses; and to policy makers, who have mental models of what policies allow. As Einstein reputedly said, "Everything should be made as simple as possible, but no simpler."
- *Least privilege.* A person should have access to only the minimum amount and type of information needed to do his or her job. For example, someone analyzing an individual's usage-pattern data should not have access to that individual's financial data. Similarly, a statistician analyzing current funds in order to forecast future funds should have access to financial information but not to customers' names and addresses.
- *Breadth.* By their nature, security policies and mechanisms need to be centrally managed, with each component of the organization managing security having a limited amount of autonomy consistent with the central policy. This approach prevents the balkanization of security but allows individual components to take local factors into account.
- *Comprehensiveness.* The principle of psychological acceptability[1] states that security mechanisms should be as unobtrusive as possible. One application of this principle is to require users to enter data only once. Then, if the data are needed multiple times (for example, for authentication across several systems), the session data should be retained so that the user need not reenter the data (for example, having a single sign-on for authentication). This principle functions with the principle of breadth, above, because local components must use security information gathered from other local components. This is possible only when a central authority controls the interpretation of the data. However, local components can decide how to use the data, provided the use is consistent with overall policy.
- *Holistic approach.* An organization may interact with many private-sector organizations (including financial institutions); with federal, state, and local agencies; and with private citizens. Issues of security, privacy, and authentication affect and are affected by all these relationships. Thus, the organization must consider not only its own systems and policies but also those outside its control with which it must interact, and the people and agencies that use its systems.

---

[1] J. Saltzer and M. Schroeder, "The Protection of Information in Computer Systems," *Proceedings of the IEEE* 63(9):1278-1308, September 1975.

This section has focused briefly on three aspects of what is commonly called security: namely, security, privacy, and authentication. Each is different, and presents different challenges. All are intertwined; none can be considered without regard to the other two, and decisions made for one affect the decisions that will have to be made for the other two. A brief summary of general security principles is provided in Box 3.3. Several themes that the SSA should continue to keep in mind when contemplating security, privacy, and authentication policies going forward are these: the larger societal, organizational, and technological contexts of which it will be a part; balancing risks with benefits and avoiding an overemphasis on risk, especially with regard to electronic services (in today's environment, lack of electronic service adoption is itself a serious risk for an agency facing serious workload and staffing challenges); and the need to be open to the evaluation and adoption of proven industry technologies and tools.

USER INTERFACE

Some of the current barriers to electronic service adoption are external to sponsoring organizations such as the SSA and/or they directly involve users. These include user concerns over privacy and security (discussed previously), the perceived convenience or inconvenience of electronic services, users' lack of knowledge about available services, the complexity and cost of providing electronic services, and lack of access to Internet-connected devices.[39] Privacy concerns may have a significant impact on overall user interfaces and access, because information protection measures being considered at state and federal levels may require users to provide identifying information such as personal identification numbers (PINs) and passwords. For example, encrypted data may require the cus-

---

[39]Michael Adler and Paul Henman, with Jackie Gulland and Sharon Gaby, *Computerisation and E-Government in Social Security: A Comparative International Study,* Washington, D.C.: IBM Center for the Business of Government, July 2005; Council for Excellence in Government, Hart-Teeter on behalf of the Council for Excellence in Government, *The New E-Government Equation: Ease, Engagement, Privacy and Protection,* Washington, D.C.: Council for Excellence in Government, April 2003, available from http://www.excelgov.org/usermedia/images/uploads/PDFs/egovpoll2003.pdf, accessed June 19, 2007; Internal Revenue Service, *Findings from the 2002 Wave of e-file Taxpayer Attitudinal Tracking Research,* 2002, available at http://www.irs.gov/taxpros/display/0,,i1%3D5%26genericId%3D10121,00.html, accessed May 23, 2002; Nielsen/NetRatings, *More Than One Third of All Online Users Log on to Government Sites,* New York: Nielsen/Net Ratings Press Release, March 17, 2003, available at http://www.nielsen-netratings.com/press.jsp?section=ne_press_releases&nav=1#2003, accessed June 19, 2007; D.M. West, "E-Government and the Transformation of Service Delivery and Citizen Attitudes," *Public Administration Review* 64(1):15-27, 2004; and Elena Larsen and Lee Rainie, *The Rise of the E-Citizen: How People Use Government Agencies' Web Sites*, Washington, D.C.: Pew Internet and American Life Project, April 3, 2002, available at http://www.pewinternet.org/reports/toc.asp?Report=57, accessed June 19, 2007.

tomer to provide identifying information not only over the Web but also in phone calls, so as to authorize the person at the call center to access the caller's data. This requirement can pose challenges to some customers who may have difficulty remembering such information. Other barriers are internal to sponsoring organizations and arise from institutional and cultural factors: lack of staff, an out-of-date technical infrastructure, competing priorities for resources, lack of institutional readiness, and lack of leadership.[40]

Fifteen years ago, the major Internet applications were e-mail, file transfer (ftp), and remote log-in (telnet). Nearly all home users accessed online services through dial-up connections. There were no electronic commerce sites. The emergence and burgeoning of the Web has clearly had a dramatic impact on all aspects of our lives, and there is every reason to believe that its impact will only increase. The majority of individuals now in their retirement years are not Web users, and usage currently decreases markedly for those over 70. However, by the time the baby boom generation has finished retiring, in 15 to 20 years, one can reasonably expect Web use by retirees to be much higher. The youngest baby boomers are now in their early 40s and have far different exposure and comfort levels with computers and electronic transactions than did average people in their 40s in the early 1980s. The oldest baby boomers have turned 60, and the majority of them are using the Internet (see below).

There always will be some individuals (of all ages) who will not use electronic services despite efforts to make these services accessible and easy to use. However, other people or institutions may be assisting users with these electronic services; thus, the SSA's electronic services are not just for use by individual clients, but also by the states, other government agencies, and employers, among others.

In predicting the demand for electronic information and services, it is important to understand the characteristics of Internet users—who will be online and how old will they be—as well as how online demographics will be changing. Key questions include these: Is it true that older people do not use the Internet? What are the projections, especially among baby boomers, for older people using the Internet? Who will use it and what

---

[40]S.H. Holden, D.F. Norris, and P.D. Fletcher, "Electronic Government at the Local Level: Progress to Date and Future Issues," *Public Performance and Management Review* 26(4):325-344, 2003; J.L. King, V. Gurbaxani, K.L. Kraemer, F.W. McFarlan, K.S. Raman, and C.S. Yap, "Institutional Factors in Information Technology Innovation," *Information Systems Research* 5(2):139-169, 1994; M.J. Moon, "The Evolution of E-Government Among Municipalities: Rhetoric or Reality?" *Public Administration Review* 62(4):424-433, 2002; and D.F. Norris and M.J. Moon, "Advancing E-Government at the Grassroots: Tortoise or Hare?" *Public Administration Review* 65(1):64-75, 2005.

will they use it to do? Are the users getting help with electronic services, and what are the motivations of those providing the help?

In a partial answer to such questions, the Pew Internet and American Life Project[41] reports the following statistics from 2004 survey data:

- As of February 2004, 22 percent of Americans 65 and over (about 8 million people) use the Internet, compared with 15 percent in 2000 and 2 percent in 1996. The trend is clear. Those who have learned how to use the Internet are expecting to use it in the future. Surprisingly, 77 percent of those with Internet experience have 4 or more years of experience.
- In the population already 65 and older, Internet use decreases steadily with age: in 2001, about 22 percent of Americans age 65 to 69 used the Internet, compared with about 15 percent of those age 70 to 74, about 8 percent of those age 75 to 80, and about 4 percent of those age 81 to 90.[42]
- In that population of Americans 65 and over (see the preceding bullet item), the number of female users equaled the number of male users. Compared with the general population, not surprisingly, these are people with higher incomes and more education. Many more of this set of users access the Internet at home over dial-up, rather than high-speed connections, compared with the general population who go online at home. Only 11 percent of African-Americans 65 and older use the Internet, although 21 percent of English-speaking Hispanics 65 and over use the Internet, the same as for Caucasians in that age range.
- Of those 65 and older who use the Internet, 94 percent use e-mail and well over half look for health or medical information. By the end of 2003, 60 percent of seniors who used the Internet had visited government Web sites, compared with 40 percent in 2000. Most of these users are online daily.
- Of more interest are the expectations and talents of the older baby boomers, at the time of the survey 50 to 58 years old; the Pew report calls this population the "Silver Tsunami." In 2004, Pew found that 62 percent of those 50 to 58 years old used the Internet, compared to 46 percent of those in the 59 to 68 year age range. Only 17 percent of those age 69 and older use the Internet. The older boomers are very much like the younger, Generation X, "eager users." They embrace e-mail, use instant messaging,

---

[41]Susannah Fox, *Older Americans and the Internet*, Pew Internet and American Life Project, Pew Research Center, Washington, D.C., March 25, 2004 (hereafter cited as Susannah Fox, *Older Americans and the Internet*), available at http://www.pewinternet.org/pdfs/PIP_Seniors_Online_2004.pdf, accessed July 11, 2007.

[42]Susannah Fox, *Wired Seniors*, Pew Internet and American Life Project, Pew Research Center, Washington, D.C., Sept. 9, 2001, available at http://www.pewinternet.org/pdfs/PIP_Wired_Seniors_Report.pdf, accessed July 11, 2007.

and often get their news online. "As Internet users in their 50s get older and retire, they are unlikely to give up their wired ways and therefore will transform the wired senior stereotype."[43]

• However, one in four Americans never uses the Internet and is not in a household with Internet access. People over 65 currently make up a large portion of this group.

This last group is sometimes referred to as the "Truly Disconnected," and its number has not changed over the past 10 years. These are the people who need agents to help them get the information they need. Even though this group may be a relatively small proportion of the general population, it represents a large number of people in absolute terms. There are barriers to these people getting the information that they need. For instance, a number of older people think that they do not need to access the Internet, since many services, such as Social Security, are available through multiple delivery channels. This is an opportunity for education and outreach. In addition, many older people do not own a computer, nor know how to use one. This number will decrease in time, but right now there is clearly a need for third parties or intermediaries to work with the people in this group to get them the information they need. This could be done through organizations similar to commercial tax preparers that help tax filers, through public servants such as reference librarians at public libraries, or through not-for-profit organizations that provide a variety of services to SSA beneficiaries. Just as the financial services institutions discussed in Chapter 2 do not remove all access channels other than electronic services, so the SSA will need to retain and support more traditional approaches to service provision.

Even if older people access Web sites on the Internet, many are difficult to use.[44] The Nielsen/Norman Group, a usability consulting firm, has declared that Web sites are twice as hard for older people in general as for younger people to use. In particular, older people have difficulty with small type sizes, low contrast, short links (the target size to be pointed at is small), and rolling pull-down menus. Furthermore, because of short-term memory challenges, if the site does not change the color of the links traversed, older people may have difficulty navigating.[45]

---

[43]Susannah Fox, *Older Americans and the Internet*, p. iii.

[44]See Amanda Lenhart, John Horrigan, Lee Rainie, Katherine Allen, Angie Boyce, Mary Madden, and Erin O'Grady, *The Ever-Shifting Internet Population: A New Look at Internet Access and the Digital Divide*, Washington, D.C.: The Pew Internet and American Life Project, April 16, 2003.

[45]*U-Group: Improving the Usability of Federal Communication Technologies*, Newsletter, Washington, D.C.: U.S. General Services Administration, September 2004.

One set of current usability standards for older people recommends the following:

- Less text (so that it is easier to remember),
- Plain language,
- 12 point sans serif font,
- Strong color contrast,
- Buttons that can be used to enlarge the type size, increase the text contrast, or read the text out loud, and
- Consistent page layout (so that people can learn where to look for various things).[46]

Not surprisingly, efforts to make a site more usable for seniors also increase the usability for all age groups.[47] The SSA will also need to keep in mind the requirements of individuals with disabilities or significant physical or cognitive impairments. Although the same principles and approaches may prove useful, ability-specific methodologies may be needed to ensure that the SSA site is as easy to use as possible for people of varying abilities. More generally, what are considered optimal Web site designs and standards will change over time. Remaining aware of best practices and conducting ongoing use testing to evaluate proposed sites and designs are necessary efforts for any institution maintaining an online presence.

ACCESS TECHNOLOGIES

A variety of technology options—cable, digital subscriber line (DSL), fiber-to-the-premises, metropolitan area wireless networks—are enabling increased broadband penetration. Increasingly functional access using handheld devices is available through 3G cellular networks. Although personal computers (whose prices, both absolute and performance-adjusted, have steadily dropped) appear likely to remain the major way that users access the Internet, the SSA, and indeed all government agencies, should expect that other technologies such as mobile phones, Internet-connected televisions, game boxes, and other "thin clients" will provide means for those who are not technologically savvy to use networked information and services. Consequently, the SSA should not continue to expect that most of its customers will be using Internet Explorer (or a similar browser) running on a desktop or laptop PC. To be effective and to meet the likely future needs of its users, the SSA should be prepared to eventually support a wide variety of browsers running on a wide variety of devices ranging

---

[46]Ibid.

[47]Janice R. Nall, Usability Solutions Group, U.S. General Services Administration.

from digital televisions to cellular phones. This is not to say that the SSA will need to be at the forefront in developing these services. Other private-sector industries will undoubtedly lead the way. The agency will need to be mindful of these and related technologies and how they are being deployed and used, and it will need to be prepared to take advantage of them for its own beneficiaries and users when appropriate.

Deep broadband penetration will provide the opportunity for the SSA's users to interact with the agency using technologies such as text chat or Voice over Internet Protocol (VoIP). For example, when filling out a form, a user might click a help link connecting the user to a live help desk using VoIP or chat (as some online retailers do today). Alternatively, a "stuck" user might click on a context-aware help link that would bring up a segment of video demonstrating how to continue filling out a troublesome part of the online application.[48]

SUMMARY

Like virtually all other organizations inside or outside the government, the SSA must carefully and regularly examine information technology trends. The agency simply could not perform its functions without the use of technology, and its current reliance on technology will inevitably grow as its workload grows and its access to support resources inevitably fails to keep pace. Thus, it is incumbent on the SSA to keep close track of changes and advances in technology. In some areas, notably security and privacy, the SSA seems to be doing a good job of tracking technology and incorporating current best practices into its processes. In other areas, notably databases, the SSA is not on par with technological advances. Ensuring that the SSA is able to keep pace with technology is not only a technological issue but is also an organizational issue. There are other organizational issues that have a direct bearing on the SSA's ability to roll out needed electronic services in coming years. These are addressed in Chapter 4.

---

[48]There are also technologies known as Rich Internet Applications, such as Flash, Asynchronous Java Script and XML (AJAX), and others that allow Web developers to develop richer client-side applications by migrating pieces of an application from the server to the client browser. See, for example, Jesse James Garrett, "AJAX: A New Approach to Web Applications," Feb. 18, 2005, available at http://www.adaptivepath.com/publications/essays/archives/000385.php, accessed June 20, 2007. The most familiar example of AJAX technology is probably Google maps (http://maps.google.com), which provides the user the ability to "scroll" the map by dragging it with the mouse. This scrolling occurs locally, with the AJAX application asynchronously requesting additional map segments and/or satellite imagery from the Google server. In addition to providing a richer user experience, applications written using AJAX technology reduce the load on the server. On the other hand, such approaches can conflict with the desire to let the user employ "thin clients" to access services.

4

Toward Organizational Transformation for Electronic Service Delivery

This study's charge is to examine the Social Security Administration's (SSA's) strategy for expanding its provision of electronic information and services. Developing a forward-looking strategy for electronic service delivery requires more than just an understanding of the organization's current technological infrastructure and capabilities. Successful service-delivery strategies are grounded in an understanding of the organization's strengths, the constraints under which it must operate, and the nature of the organizational support for (or resistance to) a new or updated vision of service delivery. As a result, an understanding of the organization's culture and its approach to both technology and service provision is also necessary.

This chapter provides a high-level assessment and set of impressions regarding the SSA's organizational culture and approaches to service provision on the basis of input that the committee received over the course of this study. The chapter also explores how the SSA's organizational culture, leadership, management choices, and policy choices may affect the agency's ability to plan and execute a new service-delivery strategy that emphasizes electronic services. It presents a set of findings and recommendations to assist the SSA in effecting a transition to the future.[1] The

---

[1]As in Chapter 3, note that this overview is of necessity brief and based on comparatively small amounts of data and input. A comprehensive assessment of such a large organization was outside the scope of this committee's activities; the committee tried to focus particularly on the capabilities and functionalities related to electronic services provision.

chapter concludes by laying out a vision of government in which information technology (IT) has transformed the service-delivery process.

ORGANIZATIONAL CULTURE AND ITS IMPLICATIONS

Organizations and individuals in organizations make decisions about the technologies that they will design and deploy for a variety of reasons, some stemming from the organizational culture. This section discusses some potential impediments in the realm of organizational culture to an effective transformation of the SSA's service-delivery strategy. To be sure, some of these potential impediments, including a cautious approach and instincts to preserve the status quo, are inherent in any large organization and especially in government organizations. Nonetheless, such impediments present a particular challenge when an organization reaches a critical juncture such as the one that faces the SSA now.

Organization and organizational culture—especially as they relate to structure, decision-making processes, approaches to problem solving, service provision, change management, and so on—are significant determiners of the success of any large organization's attempt to make fundamental and critical changes. By "cultural issues," the committee is referring to a combination of history, leadership, existing social and political relationships, overarching regulations and rules, and existing operational procedures that determine or influence the agency's activities and ways of doing business.

The experience of industries such as the financial services industry demonstrates that development, deployment, and integration of electronic services entail just such critical and fundamental change. Culture has an effect at all stages and in all components of any path forward in this area. Thus, the committee has examined relevant SSA cultural issues in order to ensure a more complete and well-rounded report and a more relevant and realistic set of recommendations.

To provide a historical context and a variety of perspectives, this section draws on presentations and discussions from meetings of the committee, the findings of previous National Research Council reports, General Accounting Office (GAO; now Government Accountability Office) reports and testimonies before Congress, and various consultants' reports. Together, these sources provide strong evidence that organizational culture has been a challenge to past attempts to improve electronic services provision at the SSA. Moreover, despite these previous findings, the organizational culture does not seem to have changed or evolved as quickly as it could (or should) have.

The SSA's Organization and Culture—Features and Issues

The SSA is a very large government institution with a long history of valuing and providing personal service to the people of the United States.[2] It has built up an impressive infrastructure over decades to support its overriding objective of providing effective personal service to its beneficiaries. It has adopted as its organizational structure a traditional hierarchical bureaucracy of the sort that most large organizations have viewed as a model of efficiency since the Industrial Revolution. However, information technology (IT) advances—of which a hallmark is the widespread availability of information at the fingertips of most employees—that have the potential for transforming work processes and service delivery were not envisioned when the industrial model was developed.

During its study and associated briefings and conversations, the committee noted that there seemed to be an understanding of the promise of an IT-enabled organizational transformation and of its potential for supporting electronic services delivery. But the promise of such a transformation seemed to be outweighed by the concern of the organization and many of its people for the impact of such a change on SSA users and beneficiaries.

The SSA is not unique in facing cultural and organizational challenges in attempting to pursue innovative initiatives, especially those relating to the implementation of electronic information and services. In 2001, major findings from studies of the U.S. government's experiences in the development of e-government examined the strong role of embeddedness and culture.[3] In that work, "embeddedness" refers to the fact that information systems are embedded or situated in organizations with complex histories, existing social and political relationships, overarching regulations and rules, and existing operational procedures. The study found that embeddedness can indeed cause cultural issues to become important factors in determining the success of an organization's efforts to innovate. Researchers in organizational theory have found that adopting innovative technologies and procedures can significantly affect an organization's culture because their adoption usually cuts across organizational bound-

---

[2]The SSA operates a vast network of some 1,500 offices and 65,000 staff distributed throughout the country (see Chapter 1). The agency's policy is to provide customers with a choice in how they conduct business with the SSA. Options include visiting or calling a field office, calling a toll-free number, contacting the SSA through the mail, or, in some cases, using Social Security Online, the agency's Web site at http://www.ssa.gov. In addition, 54 state Disability Determination Service (DDS) agencies make initial and ongoing disability determinations.

[3]Jane Fountain, *Building the Virtual State: Information Technology and Institutional Change*, Washington, D.C.: Brookings Institution Press, 2001 (hereafter cited as Jane Fountain, *Building the Virtual State*).

aries.[4] Coming to realize this can then cause the organizational culture to resist the innovations.

The diffuse nature of electronic service responsibilities across the agency and the SSA's e-government initiative can be seen as an example of innovation that generates such resistance. This innovation may be perceived by some as threatening the institutional status quo. The result seems to be difficulty in undertaking innovative strategies—that is, making significant changes, such as moving strongly to electronic services—because the organization has a strong investment in preserving its own culture (although, as is often the case, this is not necessarily a deliberate choice). Moreover, there are admittedly reasons for defending the status quo—the SSA's ultimate mission of serving its beneficiaries (and issuing millions of checks on time each year to millions of beneficiaries) has been successful. All of this suggests that the SSA's organizational culture is a factor that must be considered and dealt with if the promise of an IT-enabled organization, structured to support the kind of electronic services that are increasingly expected by society at large, is to be achieved.

These observations about the impact of the SSA's organizational culture and resistance to change become increasingly important because the technological and social environments in which the SSA operates are changing rapidly, making the alternative of an IT-enabled organizational transformation an increasingly attractive alternative for supporting effective operation and service delivery. The number of the SSA's beneficiaries will soon expand rapidly as baby boomers retire or qualify for disability. For example, since the year 2000, the number of people who file for disability increased by 500,000 a year, an increase of 25 percent.[5] At the same time, the scale and scope of the SSA's services continue to grow.

Service Culture Issues

From what the committee observed, the SSA workforce takes great pride in fulfilling its mission and has a very strong commitment to and long tradition of providing personal service tailored to the individual client. As a result, SSA devotes quite substantial resources today to its face-to-face and telephone service channels. In addition, the cultural mores within the SSA seem to equate electronic or online services with imper-

---

[4]Jane Fountain, *Building the Virtual State*. See also, for example, Lionel Pierce, "New Government: Managing the Transformation," Discussion Paper No. 20, 2004, available at http://www.agimo.gov.au/publications/2004/05/egovt_challenges/issues/transformation, accessed July 11, 2007.

[5]Mary Mosquera, "Case Files Travel Lighter, Faster," *Government Computer News*, Oct. 9, 2006, available at http://www.gcn.com/print/25_30/42177-1.html, accessed June 14, 2007.

sonal service instead of recognizing this service channel as often being more responsive in terms of cycle times, convenience, user satisfaction, burden reduction, and accuracy (see Chapter 2).

Personnel Issues

Like virtually all federal agencies, the SSA faces a projected brain drain. A substantial number of the agency's most experienced civil servants, who best understand the complex benefits-determination processes and interworkings of the outdated technology infrastructure, are eligible to retire. Thus, one immediate challenge involves human resources and workforce considerations. The agency's planning estimates for the workforce indicate that just over 40 percent of the workforce will be retiring by 2014, just when the baby boom generation approaches its peak disability and retirement years. Consequently, the SSA acknowledges that a significant human capital challenge is to develop strategies to maintain a capable workforce to handle the workload increase, despite the retirement wave. To address this challenge, the SSA reports that it developed its first human capital plan in 2004, updating the plan in 2005, along with completion of its future workforce transition plan. The latter lays out the SSA's plans to recruit, hire, develop, and retain a diverse workforce. The agency reports having hired more than 18,000 permanent employees since 2001.[6] The SSA also reports that it is focused on workforce development and retention, especially for employees providing services to the public, and that it is working to improve training and development programs, including online training.[7]

The committee notes that the start of the SSA's hiring wave apparently preceded completion of the 2004 Human Capital Plan. This timing raises possible concerns about the skill mix of the newly hired workforce (see the subsection below entitled "Strategic Service-Delivery Plan"). An updated employee skill mix will be needed to support technological innovations indicated as part of a modern electronic services strategy. More important perhaps is the need for new employees to bring to the SSA a fresh outlook on IT and electronic services. The SSA may find it challenging to establish a more modern employee skill mix, and the cadres of new employees who will embrace and lead a cultural transformation may encounter resistance.

---

[6]Testimony of Jo Anne B. Barnhart, Commissioner of Social Security, before the Senate Finance Committee, March 14, 2006, available at http://www.ssa.gov/legislation/testimony_031406.html, accessed June 9, 2006.

[7]Social Security Administration, *Results at the Social Security Administration: Getting It Done*, 2005, available at http://www.ssa.gov/performance/results/results2005.pdf, accessed June 9, 2006.

Finally, it is worth underscoring that aside from the transformational issues, the SSA will continue to need to recruit and retain in-house technical expertise. Government agencies generally tend to outsource a large portion of their technical work and thus run the risk of being overly dependent on their vendors. A sufficient cadre of in-house technical experts is critical in any organization to evaluating the quality of the technology solutions that are offered.

Organizational Structural Issues

Although SSA programs cross organizational lines, the committee notes that the SSA, like many government agencies and other long-lived organizations, is a heavily stovepiped organization. As described in the discussion in Chapter 1 about the SSA's current organizational structure, key elements for the development and implementation of e-government by the SSA are spread across three Deputy Commissioners' domains—Operations, Systems, and Disability and Income Security Programs. The organizational lead for e-government is a subcomponent of Operations. Moreover, three additional units play a role in e-service initiatives—(1) the Office of the Chief Information Officer, which appears to have responsibility for IT strategy and for compliance with legislation, Office of Management and Budget (OMB) directives, and Government Accountability Office (GAO) guidance concerning IT capital and investment control, and which also works with other agencies on government-wide projects such as e-government; (2) the Office of Electronic Communications in the Office of the Deputy Commissioner, Communications, which is responsible for online informational (as opposed to transactional) content; and (3) the Office of the Chief Strategic Officer, which is responsible for strategic planning.[8]

As in any large organization, this stovepiping generally tends to make cross-program communications difficult, and it can greatly complicate the kind of coordination needed to ensure success in interorganizational initiatives. In particular, such dispersed responsibilities and the relatively low-level lead for e-government initiatives make success in creating an ambitious spectrum of new e-government initiatives particularly challenging. Of particular note is the effective split between the electronic services organization's mission and that of the communications

---

[8]This description is current at the time of this writing and the discussion is based on material from the SSA's public Web site, "Organizational Structure of the Social Security Administration." See http://www.ssa.gov/org/ssaorg.htm, accessed June 9, 2006, and May 30, 2007 (for up-to-date organizational information).

organization—which effectively divide responsibility and authority for the Internet/electronic face of the SSA.

Given the SSA's current stage of maturity with respect to electronic services (see Chapter 2), raising the level and authority of the electronic services organization could be helpful for the organization to successfully address its challenges by making electronic services a more prominent part of its service-delivery plans. The alternative is to maintain the distributed nature of electronic services and have responsibility for various facets spread across multiple Deputy Commissioners' domains. Although analogous experiences in the private-sector financial services industry (see Chapter 2) indicate that resistance to such changes can be expected, those experiences also suggest that these changes are important, both to support the move to electronic services and to precipitate needed cultural change.

Risk Management in the Federal Context

Being subject to constant and intense scrutiny from Congress and various federal oversight agencies understandably influences the SSA to be reluctant to assume risk. Yet failing to make needed changes can incur costs that can eventually outweigh the risks incurred in making such changes. There is a larger need to institutionalize effective risk management in order to make increasingly accurate assessments of both the upside and the downside risks from decisions both to make and to defer change.

A 1990 report of the National Research Council on the SSA's systems modernization strategy observed that "the SSA's intrinsic problem is one compounded of mammoth size, an external expectation for inherent agility to accept change, a built-in inertia that impedes change, often late-breaking changes, and all overlaid with a demand for maximum accuracy and stable operations. So far as the nation is concerned, the SSA is a big information production line that must deliver checks to its clientele consistently and with certainty."[9]

The desire to avoid "more bad publicity" is a factor that seems to have significantly reinforced a cautious organizational culture. In 1997, the Personal Earnings and Benefit Estimate Statement (PEBES) situation (also mentioned briefly in Chapter 2) garnered the SSA significant negative attention. Privacy concerns among the public and among privacy advocates regarding the SSA's implementation of Internet-based access

[9]National Research Council, *Systems Modernization and the Strategic Plans of the Social Security Administration*, Board on Telecommunications and Computer Applications, Washington, D.C.: National Academy Press, 1990.

to individuals' PEBES information gave rise to front-page news stories[10] and congressional hearings.[11] The SSA Commissioner quickly suspended the service.[12]

The PEBES experience illustrates the kind of attention that can be brought to bear on agencies such as the SSA when they implement new technologies. Although such experiences can impose costs and increase wariness regarding attempts at other kinds of changes, it is also important to assess what positive lessons can be learned. For instance, if the SSA had vetted its plans for PEBES more publicly beforehand, it might well have been able to head off some of the negative effects—by, for example, revising its authentication solution or simply by not surprising the privacy community. Such consultation prior to the launch of the new service might also have given the SSA an opportunity to present the risk-reward trade-offs before the press presented only the downside risk. Articulating the benefits of future improvements and securing the support of a broad range of users (including beneficiaries, third parties, and partners) is one way to ensure public support in case there is a mix of opinions that otherwise might derail a worthwhile service offering.

A final lesson about the PEBES episode is that too much should not be made of it. Today, people are much more comfortable about (secured) online access to private information, and much more is known about how to implement online authentication.

Finding: The SSA may be missing important opportunities to make sustained improvements in its service delivery because of an overemphasis on the potential risks of modernizing its service-delivery strategy and a lack of emphasis on the long-term risks associated with *not* revamping that strategy.

Recommendation: When evaluating new electronic service-delivery initiatives, the SSA should when appropriate seek to balance risks and rewards by recognizing such upside benefits from automation as cost reduction, fraud prevention, and customer satisfaction.

---

[10]See, for example, "Few Key Bits of Info Open Social Security Records," *USA Today*, April 7, 1997, p. 1.

[11]U.S. Congress, Subcommittee on Social Security, House Committee on Ways and Means, Hearing on the Social Security Administration and the PEBES Program, May 6, 1997.

[12]PEBES went online in March 1997. The Commissioner suspended online receipt of PEBES data on April 9, 1997. See U.S. General Accounting Office, *Social Security Administration: Internet Access to Personal Earnings and Benefits Information*, Testimony Before the Subcommittee on Social Security, Committee on Ways and Means, U.S. House of Representatives, GAO/T-AIMD/HEHS-97-123, Washington, D.C., May 6, 1997.

Strategic Service-Delivery Plan

The GAO made recommendations in 1993 and again in 2000[13] that the SSA complete a service-delivery plan to ensure that its human capital and other key investments are put to the best use. In 1998, the agency took a first step by beginning a multiyear project to monitor and measure the needs, expectations, priorities, and satisfaction of customer groups, major stakeholders, and its workforce. In 2000, the SSA took another step by completing a document that articulated its vision for how the agency would function in the future.[14]

However, the overall impression of the committee is that the SSA has been slow to develop and implement strategic service-delivery plans, despite repeatedly being encouraged to do so by outside auditors and experts. The reasons for this are not entirely clear, although cultural issues, sometimes driven by external factors,[15] are likely at least partially responsible. Contributing factors may be a lack of sustained leadership commitment over the years and an organizational culture that appears to equate electronic or online services with impersonal rather than more-responsive service. Here too, these cultural predispositions need to be overcome in order to enable the SSA to embrace the proactive planning needed as the basis for electronic service-delivery initiatives.

As the committee noted above in the discussion of the retirement and hiring waves, the SSA sometimes takes actions with enduring consequences in advance of a strategic plan for action.[16] This way of proceeding has led to outside criticisms of the SSA's workforce and technology initiatives as being developed out of sequence and lacking a cohesive

---

[13]See U.S. General Accounting Office, Testimony Before the Subcommittees on Human Resources and Social Security, Committee on Ways and Means, House of Representatives, *SSA Customer Service: Broad Service Delivery Plan Needed to Address Future Challenges,* Feb. 11, 2000; and U.S. General Accounting Office, Testimony Before the Subcommittee on Social Security, Committee on Ways and Means, House of Representatives, *Social Security Administration: SSA Needs to Act Now to Assure World-Class Service,* T-HRD-94-46, Washington, D.C., Oct. 28, 1993.

[14]Social Security Administration, *Social Security 2010 Vision,* Office of the Commissioner, SSA Pub. No. 01-016, September 2000.

[15]Such factors include everything from congressional requirements, to executive branch mandates, to employee resistance to reorganization, and so on. The committee does not seek to downplay these factors in any way, but making recommendations to these organizations is outside the scope of this report.

[16]See U.S. General Accounting Office, *GAO Report Performance and Accountability Series: Major Management Challenges and Program Risks, Social Security Administration,* Washington, D.C., January 2003. GAO found that the SSA was taking human capital measures in the absence of a concrete service-delivery plan to help guide its investments.

strategic vision.[17] Indeed, the committee's impression is that the SSA's de facto service-delivery strategy appears to be oriented toward preserving existing organizational structures and mores instead of responding to emerging technological capabilities and shifting user demands for more convenient service channels.

The GAO also criticized what it described as the SSA's lack of a broad vision for customer service, noting that "this broad vision, as well as a more detailed plan spelling out who in the future will be providing what service and where, is needed to help the agency focus its efforts to meet its future challenges."[18] As is expanded on later in this chapter, this committee believes that a very broad new vision of the seamless delivery of SSA information and services through different (but predominantly electronic) service media centered on state-of-the-art technology is indicated. A thorough and articulate plan is needed as the basis for proceeding toward such a vision. At the same time, the committee recognizes the difficulties that federal agencies face in projecting programmatic cost savings resulting from the use of new technologies and methods and then having to live with the consequences of those projections (e.g., budget cuts) whether or not the savings are realized.

Finding: The SSA's organizational structure does not support the establishment of a strategic focus in electronic services that is sufficiently high-level and broad-based. The SSA has an opportunity to be more proactive in fundamentally reassessing its customer service value chain and, for as many customers as possible, focusing on the potential substitution of electronic services for other delivery channels, such as paper mail and face-to-face interactions in field offices.

Recommendation: The SSA should make an unambiguous, strategic commitment to electronic services as part of its long-term service-delivery strategy, placing a central emphasis on electronic services that encompass timely and up-to-date information for users, partners, and beneficiaries.

---

[17] U.S. General Accounting Office, Testimony Before the Subcommittees on Human Resources and Social Security, Committee on Ways and Means, House of Representatives, *SSA Customer Service: Broad Service Delivery Plan Needed to Address Future Challenges*, Washington, D.C., Feb. 11, 2000.

[18] Ibid.

OPPORTUNITIES FOR CHANGE

Several key issues were largely unaddressed in the briefings and documents made available to the committee: (1) what kinds of metrics and measures the agency will use to monitor and assess its efforts over time, (2) what kinds of marketing and promotion of services it will undertake in order to encourage the use of electronic services, and (3) how the SSA's core competencies and potential partnerships could be combined to strengthen and enhance service provision across the board. The subsections that follow examine each of these questions in turn.

Metrics and Measures

The National Research Council's 1990 report entitled *Systems Modernization and the Strategic Plans of the Social Security Administration* recommended that the SSA develop and use suitable performance metrics and measures in its management process:

- As part of its recommendation that the SSA adopt improved IT management processes, that study committee urged the SSA to "thoroughly forecast and justify expansion and upgrade of the existing information systems infrastructure on a continuing basis; *quantify and define in advance the specific performance goals to be provided its clients; measure and monitor actual performance* [emphasis added]; and plan for the orderly introduction of new services consistent with available personnel and budgetary resources."
- With respect to service quality, the same committee recommended that "the Social Security Administration quantify each aspect of service quality (e.g., elapsed time to complete), monitor its overall performance, and *manage against such a priori service goals* [emphasis added]."[19]

This committee concurs with the above conclusions. The establishment of metrics to evaluate the effectiveness of various services and delivery channels is an important component of a service-delivery plan. Such metrics are complemented by the establishment of specific, agreed-on operational approaches to measurement. Although a typical cost measure is dollars, benefit measures may be difficult to monetize. Careful discussion that involve user communities (such as beneficiaries, third parties, partners, and SSA employees), can be useful in coming to consensus about

---

[19]National Research Council, *Systems Modernization and the Strategic Plans of the Social Security Administration*, Board on Telecommunications and Computer Applications, Washington, D.C.: National Academy Press, 1990, p. 4.

effectiveness measures, and such discussions themselves are often what is most useful to an organization.

As a general rule, cost-reduction and efficiency measures should be balanced by effectiveness metrics that reflect operational objectives. While cost is easily measured, defining metrics of effectiveness is a significant challenge. Many enterprises wrestle with this issue since management frequently sees IT as a cost center, thus emphasizing cost reduction, and does not tie the operational effectiveness of its lines of business directly to IT capabilities. For example, effective risk management relies on effectiveness in quantifying the costs and benefits from changes and decisions that are under consideration. Metrics of this type are generally the basis for the orderly collection of measurements of specific current and proposed costs and benefits that can then be the basis for sound decision making and orderly change. The committee did not receive information regarding SSA metrics of this type.

In addition, when any organization is working in a metrics-oriented fashion, effectiveness metrics against which costs can be compared should be established and used as vehicles for project management and control before major sums of money are committed to projects. That is, *management by cost-effectiveness measurement* should be established. The committee did not receive information from the SSA about the use of business case analyses to determine when projects are succeeding, when they are not, and what might be done about ineffective projects.

Measuring cost involves more than simply estimating the dollar costs of carrying out a project. In some cases, for example, there may be substantial costs in *not* proceeding with a project. These costs might include increased risk of catastrophic consequences from failing to proceed with replacement of critical infrastructure. Effective measures of opportunity cost must take such issues into account. Chapter 3 indicates that the SSA seems to the committee to be incurring serious ongoing risk in continuing to base much of its operational capability on the Master Data Access Method (MADAM) system. Although the cost of replacing MADAM may be considerable, failing to replace it seems to increase the risk that MADAM may become (perhaps quite suddenly) unsustainable, rendering many key SSA operations problematic or impossible. The cost and impact on the agency and public when MADAM can no longer be supported will be enormous—the committee believes that this is a "when" not an "if" proposition. Determining the cost-effectiveness of a project like MADAM conversion should be based on a measure of cost that takes all of these issues accurately into account. Such a determination should also consider likely increases in electronic services adoption due to the improved functionality that results from using an up-to-date relational database technology.

Another important area for measurement is the return on investments in security, authentication, and privacy protections. Although this is generally recognized as a problem (not just for the SSA), there has been some success in developing approaches for measuring the effectiveness of an organization's security, privacy, and authentication efforts.[20] Any organization needs to determine how to measure the effectiveness of its security, privacy, and authentication efforts.[21] Once determined, such metrics can help the organization understand how well it is doing with its efforts to keep its data and resources secure, how well it is protecting the privacy of its customers, and how effective its authentication methods are. These metrics can also allow it to show others, including its customers and interested parties (such as lawmakers, in the case of the SSA), how well its efforts are working. Implicit in this is that the metrics are meaningful and measure qualities of concern; determining what to assess and devising those metrics and corresponding measures are an important challenge for the organization.

In general, metrics and measures should stem from a careful examination—by tiger teams, internal work groups, and so on—of goals that are ultimately quantified. The units of this quantification would be used to identify the metrics. Implementing data-collection processes in the framework of those metrics would yield measures. Engaging in dialogue with managers from other organizations that manage in a metrics-oriented fashion and drawing from their "lessons learned" may be useful. In some cases, organizations find it helpful to engage the services of management consultants and/or to have ongoing dialogues with people who have implemented this approach in related institutions, such as large banks or brokerages.

The establishment of quantitative goals is useful for more than just having the basis for deciding whether or not to undertake a project and for determining the degree of success of a project. The process of establishing such quantitative goals also serves the useful purpose of causing interested organizational entities to think carefully and specifically about

---

[20]One discussion of measures of effectiveness for security efforts is available in the *2006 National Infrastructure Protection Plan*, published by the Financial Services Sector Coordinating Council, available at https://www.fsscc.org/reports/2006/NIPP_Plan.pdf, accessed July 11, 2007.

[21]Two types of measures seem appropriate—these are just illustrative examples: The first type is *development measures*, indicating how closely the design and implementation are believed to meet the policies. This is essentially a function of the development process, similar to high-assurance engineering methodologies. The second type is *testing measures*, indicating how well the systems meet the policies in practice. For example, penetration studies or technical audits fall into this class. This type of measure is necessary because in security generally, errors in the implementation, deployment, and management of systems will not be detected, let alone analyzed, during the development phase.

precisely what the goals of a project should be, which capabilities it should have and which are of less importance. If there is discussion or dispute about the inclusion or degree of thoroughness of specific capabilities, agreement about metrics and measures can help determine whether or not specific capabilities are likely to be sufficiently useful to justify their costs. Thus, the establishment of cost and effectiveness metrics and measures can be the basis for more effectiveness in ongoing project management, as well as for more effectiveness in determining overall project success. Such consultations will help build support for proposed investments in service delivery both inside and outside the organization. At the same time, when exploring cost and effective measures, the uncertainties of new approaches should be kept in mind. Where savings and efficiencies might result from the deployment of electronic services, the committee is mindful of the risk that budgets may be reduced accordingly (sometimes even if the anticipated efficiencies are not realized). When efficiencies are gained, there is an opportunity to redeploy those resources elsewhere as part of the SSA's broad service-delivery strategy or to manage other critical workloads. The committee recognizes and is sympathetic to this challenge for the agency.

The SSA should pay significant attention to the identification of appropriate metrics and the continual gathering of adequate measurements needed to ensure effective decision making. The agency should place major emphasis on the systematic reduction of its organizational goals to metrics that are quantifiable, although not necessarily monetized. As noted earlier, typical areas where electronic services might be compared with other service channels include customer satisfaction, cycle time, error rates, cost per transaction, paperwork burden reduction, and even impact on program integrity (that is, fraud rates). The SSA should consider putting in place organizations and processes that use such metrics to gather measures of its performance. The metrics and measures should be the key components in a program of continuous improvement, aimed at demonstrating that the SSA continues to do a better job of meeting the various needs of its diverse publics and user communities.

Finding: The establishment of appropriate metrics and measures to evaluate the effectiveness of various services and delivery channels is an important component of an effective service-delivery plan.

The committee understands that it seems easy enough to simply assert the importance of quantitative measures in managing toward meeting goals, but that putting in place the details of how to do so may be far more difficult. However, the establishment of effective, agreed-on metrics, complemented by agreed-on measures, would provide immediate benefits for the better management of projects, and potentially, they could better justify requests for funding for the deployment of electronic services.

Recommendation: The SSA should define and use metrics and measures to assess and improve its service delivery across all channels, including electronic services.

Marketing and Promotion of Electronic Services

The SSA has historically done a great deal of marketing, outreach, and providing of public information and education, using major media (print or broadcast) outlets. Moreover, it has used private marketing firms, the Ad Council, and major polling organizations to assist in the development of media campaigns. Since the committee received a comparatively brief presentation on the SSA's marketing strategy, it did not have enough information to assess whether or not the SSA has translated a comprehensive understanding of its diverse users' needs into an effective marketing campaign for electronic services, or whether it has chosen to promote electronic services over other service-delivery channels for certain categories of users.[22]

In some ways, this approach is consistent with much public-sector marketing, which is more likely to take the form of "informing and educating" the public instead of encouraging them to use one product (such as electronic services) over another (such as paper or in-person services). Generally, government agencies do not have marketing expertise that is comparable with what might be found in a product-oriented consumer business. As a result, to market a new service or service channel aggressively, agencies may need to obtain outside assistance and use the most effective media outlets (including prime-time television). Further, since it is a cultural change for some agencies to think of the public and citizens as "customers" in a business context, there may be institutional reluctance to do so. Complicating matters, commercial and business marketing models do not always translate fully to government contexts, given the different constraints and expectations that government agencies face.

Although marketing and promoting electronic services might be a challenge, at least one federal agency has proven that it is possible and can be successful. After the passage of the Restructuring and Reform Act of 1998 (RRA '98)[23] and the creation of an organization devoted solely to electronic tax administration (ETA), the Internal Revenue Service

---

[22]"Marketing" is purposeful, goal directed, and usually quantitatively driven. Marketing a service is not synonymous with informing and educating potential users.

[23]Public Law 105-206, 112 Stat. 685.

(IRS) launched a substantial market-research and promotional campaign in 1999. The committee received a briefing from a former IRS Director of Electronic Tax Administration; additionally, one committee member served as a member of the IRS executive team that led that effort.

The IRS began the campaign by contracting with a "Madison Avenue" advertising firm and then investing in objective market research that identified reasons for e-file adoption and nonadoption and in market segmentation so that the IRS knew the demographic characteristics of adopters and nonadopters. It then "rebranded" electronic filing (to IRS e-file)—based on the market research—as a prelude to both public service announcements and paid placements for advertising. The IRS worked with its tax-preparation and related software-company partners to coordinate marketing messages for the IRS e-file brand that the public and private sector both used. As an example, firms authorized by the IRS to submit e-filed returns to the IRS could use the e-file logo on their advertising and product packaging. In a marked departure from public-sector "inform and educate" efforts, the IRS used proven private-sector marketing techniques to promote e-file. Most important for the SSA to consider though, the IRS was able to find for-profit partners in the tax-preparation industry that invested their marketing budgets and talents to promote e-file to the public, to the mutual benefit of the IRS and the partners. Public adoption of e-file exceeds 50 percent of all individual tax returns, in part because the IRS and its commercial partners became more aggressive in the marketing of e-file.

The previous discussion of metrics and measures applies to advertising and marketing as well: quantitative goals can be used to determine whether a particular campaign is succeeding, or whether it should be reexamined and perhaps replaced. For example, the IRS uses specific quantitative goals (such as repeat customers, brand recognition, customer satisfaction) to guide its marketing activities.[24] The costs of such marketing activities are to be at least recouped by savings obtained from the greater efficiency of dealing with electronic tax returns as opposed to paper. The committee did not receive information about similar use of quantitative metrics of this type to guide the SSA's management of electronic services marketing.

Core Competencies and External Partnering

In an increasingly complex and demanding world, many organizations, including government agencies such as the SSA, are constantly

---

[24]Stephen H. Holden, *A Model for Increasing Innovation Adoption: Lessons Learned from the IRS E-File Program*, Washington, D.C.: IBM Center for the Business of Government, 2006.

being pressed to provide more—and always more challenging—services. Many organizations have chosen to outsource functional jobs that they had previously supported in-house, such as food service, payroll, travel services, duplication, and so on, thus freeing the organizations to concentrate only on those "core competencies" that they feel they must maintain internally for their survival. Organizations have often sought to concentrate on what is deemed essential and to devote their efforts to being highly competent at what really matters, while relegating support for peripheral capabilities to specialist business partners. Many organizations have found that the outsourcing of peripheral task areas can strengthen their ability to perform core functions.

The SSA is no exception to this way of working and has outsourced many functions over the years. At the same time, it can sometimes be a challenge for an organization to determine what really matters and what is peripheral, and it can be difficult to recover from mistakes in the determination. Given the ongoing transition to more electronic services and the emerging demand from a variety of user communities for large-scale and highly effective electronic services, the SSA will face questions about which functions to maintain and which to outsource or partner with others for.

In common with other federal agencies, the SSA must address privacy issues as it considers expanded outsourcing. Among other requirements, the E-Government Act of 2002 (Public Law 107-347) requires privacy impact assessments for agency or contractor systems that maintain personally identifiable data about members of the public.[25] Moreover, outsourcing decisions must be always open to reconsideration. The need for continuing discussions of this issue is another challenge that the agency must face. In the area of electronic services, organizations often find it useful to consider what their core competencies need to be and then to seek ways to partner effectively with others to fulfill other functions.

There are precedents for partnering in the federal arena. In some cases, legislative changes were sought and enacted; in others, an agency was able to develop partnerships within existing law and regulation. For example, the IRS and the Department of the Treasury believed it necessary to obtain legislative authority to accept credit card payments for income taxes because of the financing implications of merchant fees. Such authority was obtained through the IRS Reform and Restructuring Act of 1998. The IRS implemented the "Free File" tax program with private-sector

---

[25]See Joshua Bolton, "OMB Guidance for Implementing the Privacy Provisions of the E-Government Act of 2002," M-03-22, Sept. 26, 2003, available at http://www.whitehouse.gov/omb/memoranda/m03-22.html, accessed July 11, 2007. OMB requires the privacy impact assessment to be submitted with OMB Exhibit 300.

partners under explicit and implicit existing authority. The IRS Reform and Restructuring Commission that led to RRA '98 did include language supporting private-sector partnerships; IRS e-filing staff and the private sector worked together to help develop the relevant language for both credit card payments and Free File.

In the case of the SSA, the agency may have some flexibility to explore partnerships under its current legislative and regulatory framework. New authority may be required for some types of partnerships. The committee suggests that, in exploring possibilities for partnerships, the SSA use an open process that involves beneficiaries, third parties, and other user communities, as well as Congress. Productive partnerships might be possible in areas such as partnering with the states to enable the delivery of certified, electronic vital records evidence (e.g., copies of birth certificates or certified dates of birth, copies of death certificates or certified dates of death, copies of marriage certificates or certified marriage information) to streamline claims preparation and processing. For example, the SSA's current online application process for retirement benefits requires the applicant to send a certified paper copy of his or her birth certificate to the SSA or bring it to a field office. If applicants were able to include identifying information about their birth certificates (or circumstances of birth) in the electronic application sufficient for use by the SSA to obtain certified birth information through a partnership with the states, then the application process could be entirely electronic, faster, and would not generate visits to local SSA offices. Another type of partnership would be partnering with large payroll-processing services to increase the adoption of electronic wage reporting.

Cultivation of effective interactions that take advantage of electronic services where possible with a broad range of user communities could deliver substantial benefits to beneficiaries and the SSA as well as to third parties. As one example, the SSA has an opportunity to promote good standards work in a variety of areas such as Web standards, data exchange standards, document standards, and so on, through the publication of well-defined requirements and active participation in industry bodies. The SSA, by dint of its size and importance, can give tremendous momentum to selected standardization efforts: for example, the agency could have a tremendous positive influence on the evolution of data interchange standards in its area of the '"information ecosystem" if it (1) knows its own needs and (2) makes an active effort to provide leadership to the community. Its scale and resources make it an ideal leader in a number of areas if it can develop the focus and will to do so.

On another front, the committee received briefings indicating that user groups such as third-party representatives or not-for-profit special-needs advocacy groups often are willing to collaborate more closely with the SSA, thereby potentially minimizing the impact of part of the agency's increasing workload. Some reluctance on the part of the SSA to do so is understandable in that aspects of some SSA partnership relationships have attracted attention from watchdog groups and others (for example, when the SSA's use of personal information obtained from resellers did not adhere to Fair Information Practices[26]).

Another type of partnership might involve increased cooperation with third parties for filing claims or providing advice to beneficiaries regarding the implications of SSA rules and regulations. For example, the SSA might establish online partnerships with third parties that would file disability claims on behalf of their customers. It seems quite reasonable that the SSA can and should be the only organization that can receive, process, and validate benefits requests. But that does not imply that only the SSA can counsel and assist benefits applicants regarding the application process. The committee heard in briefings that there is a community of service providers that is eager to assist in the preparation of claims and benefits requests. If the SSA were actively to solicit and promote the establishment and growth of this community of third-party providers, it could have the immediate benefit of reducing the SSA's current workload, helping deal with resource scarcities within the SSA. It could also free the SSA to pay greater attention to such core competencies as the adjudication of claims and the clarification and streamlining of regulations and procedures that ultimately govern decisions about benefits eligibility.

When considering the design and functionality requirements of services made available electronically, the SSA should take steps to enable third parties to assist the SSA in helping their joint user base. To further facilitate effective relationships, the SSA should standardize and enforce improved interfaces to both its internal and external systems. The agency's workload might also be reduced if other parties and agencies (with appropriate authorization) could gain expedited access to SSA systems and data—for example, giving priority service to entities that drive a significant volume of online transactions.[27] In addition, there may be legitimate research questions for the social sciences or other areas that

---

[26]See U.S. Government Accountability Office, *Personal Information: Agency and Reseller Adherence to Privacy Principles*, GAO-06-421, Washington, D.C., April 2006. The SSA noted in its comments that it had instituted internal controls to prevent improper disclosure and was exploring options for enhancing its policies and internal controls over information from resellers.

[27]For example, the IRS Restructuring and Reform Commission that led to the Restructuring and Reform Act of 1998 (RRA '98) included language supporting IRS efforts to work with the private sector to increase adoption of e-filing for tax returns. The Internal Revenue Service, the Department of the Treasury, and the E-File program all agreed on the need to increase public adoption and were able to build a consensus for a number of legislative changes to support public adoption of e-file. As an example, RRA '98 included authority for the IRS to accept credit card payments and electronic signatures as a way to make e-file completely paperless, reducing costs for the IRS and making it more convenient for taxpayers.

would benefit from access to various types of SSA data. While privacy and security concerns—not to mention legislative constraints—must certainly dictate restrictions on such access, there may be opportunities for the SSA to contribute to and learn from these research areas.

When it comes to cultivating partnerships, the SSA's location in Baltimore, Maryland, is both a real and a perceived barrier to interacting more routinely with staff of other government agencies at the federal level. Both formal and informal, face-to-face interactions are valuable in initiating and sustaining human networking and professional relationships. Although the SSA is usually involved in formal interagency meetings and activities, the committee believes that the agency's geographic distance from the nation's capital likely discourages the informal interactions that take place routinely in metropolitan Washington, D.C., and may also discourage the movement in terms of employment of skilled professionals between the SSA and other agencies, which further impedes opportunities for informal contacts and discussions. The committee believes that this comparative lack of informal interaction and engagement with outside perspectives, both within and external to government, may contribute to an inadvertently inward-focused culture.

Given these geographic constraints and a culture that focuses on personal customer service, trying to institute an electronic services strategy dependent on the effective use of IT and an expanded understanding of third parties could be seen by some in the organization as antithetical to providing personalized customer service. Further, the use of third parties is also antithetical to the hands-on customer service that SSA staffers want to provide.[28] Thus, although it may not be an agency-wide policy that third parties are not to be dealt with, the prevailing organizational culture results in some field offices having a different view on this matter.

An inward-focused culture also limits or narrows the view on how partnerships might make the SSA's mission easier or more efficient to accomplish. This concept of public and private partnerships to deliver public services is becoming known as "networked government."[29] One impact of technology is its leveling effect on organizations and the emergence of value networks. The Australian Management Advisory Committee observed that "most whole-of-government priorities require close cooperation with external groups, such as community organizations, businesses and other jurisdictions."[30] Technology accelerates this need

---

[28]This is exemplified by one panelist's story of being rebuffed by staff at an SSA field office because he was *representing* a client (and was not a client himself).

[29]Stephen Goldsmith and William Eggers, *Governing by Network*, Washington, D.C.: Brookings Institution Press, 2004.

[30]See the summary of *Connecting Government: Whole of Government Responses to Australia's Priority Challenges*, Australian Management Advisory Committee, April 20, 2004, available at http://www.apsc.gov.au/mac/connectinggovernment.htm.

because it facilitates rapid, peer-to-peer (or partner-to-partner) communication. The role of the private sector in government service delivery—and, indeed, policy development—is often overlooked or underestimated. Some e-government researchers suggest that "although the public may traditionally have thought of government as synonymous with bureaucracy, in the future government will be highly networked and delivering outcomes through federations of organizations and agencies."[31]

Finding: There are opportunities for the SSA to partner with other agencies and third parties in ways that could provide mutual benefit.

Recommendation: The SSA should undertake to understand the identities, needs, and attitudes of its various user communities and should use that information to establish effective relationships and ongoing interactions with users, potential partners, and third parties. The SSA should explore partnering opportunities and identify the changes and initiatives that are necessary in order for it to enable appropriate interaction and cross-functionality with strategic partners and to support the exchange of data with other government agencies (both federal and state) while ensuring that appropriate security and privacy measures are in place.

Achieving effective partnerships and highly networked government will involve sustained market research in which the SSA asks its users, both directly and through third-party organizations, what their needs and capabilities are for electronic services. Further, a well-defined set of user types (see Box 1.2 in Chapter 1 for some preliminary examples) used at all levels of the agency could be particularly effective in projecting a clear image of these user groups, both inside and outside the SSA. Having developed clear pictures of its various user communities, the agency should then create structures, processes, and mechanisms for entering into effective ongoing relations with these groups.

Sustained Leadership Commitment Needed

With or without specific intent, delegating responsibility without authority hinders effectiveness and tends to preserve the status quo. For a revamped service-delivery strategy to work, it will need to be guided and enforced from the very top of the organization as the preferred way to conduct the SSA's business. It is surprising that the SSA's e-government

[31] John Halligan and Trevor Moore, "Future Challenges for E-Government," Monograph, available at http://www.agimo.gov.au/publications/2004/05/egovt_challenges/overview, accessed July 11, 2007.

strategy office is not located higher in the organization. Given its current placement, it does not have the authority to create and implement its strategy even if that strategy were complete, appropriate, effective, and provided a path towards transformation. (See the recommendation on this point in the subsection above entitled "Strategic Service-Delivery Plan.")

In a report issued in 2000, the GAO stated: "The SSA is one of only a few federal government agencies with which most American families will have regular contact. Because of SSA's broad reach, the quality of its customer service can affect the public's view of government overall, and SSA has committed itself to providing world-class service to the American public."[32] The report concluded that the SSA would be challenged to maintain a high level of service to the public because demand for services is expected to grow significantly, while the expectations and needs of the SSA's customers are changing. The GAO further noted that some are expecting faster, more convenient service, while others, such as non-English speakers and the large population of beneficiaries with mental impairments, may require additional assistance from staff with more diverse skills.[33] Chapter 2 in this report provides a general overview of the kinds of functionality that world-class financial institutions routinely provide these days and observes that the users of the SSA's electronic services are likely to expect such things, and more so as time passes.

Why has effective and comprehensive transformation not yet happened within the SSA? One possibility is that the agency has lacked sustained leadership commitment at the highest levels to provide the vision and incentives to make recommendations become a reality. Another contributing factor could be that the SSA has an organizational culture and exists within a political environment that encourages a conservative and incremental approach to change. Taking risks, though, is inherent in any core business process reengineering or transformational efforts. While the SSA is committed to providing world-class service to the U.S. public—an attitude that seems, encouragingly, to permeate all levels of the agency—it may lack the wherewithal to effect the sort of changes that are needed as efficiently as they are needed, in part because of its organizational culture. Without strong leadership, an innovative vision, and the power to carry out that vision, change will continue to be slow and likely will preserve the existing organizational status quo.

---

[32]U.S. General Accounting Office, Testimony Before the Subcommittees on Human Resources and Social Security, Committee on Ways and Means, U.S. House of Representatives, *SSA Customer Service: Broad Service Delivery Plan Needed to Address Future Challenges*, Washington, D.C., Feb. 11, 2000.

[33]Ibid.

Thus, until electronic service delivery becomes culturally and organizationally ingrained across all program lines as desirable for users and needed to preserve organizational knowledge resources, the responsibility for planning and executing electronic information and service delivery should be elevated in the SSA. Note that this proposal is not in any way calling for a substantial reorganization. This elevated role for electronic services would serve as the organizational focal point for conducting market research, establishing and maintaining partnerships, setting priorities for the development of new electronic information and service-delivery systems, and working with the IT organization to ensure that appropriate technologies are put in place to support electronic services. Over time it is likely that some of these responsibilities could devolve to program areas in a way that is analogous to what has taken place in the private financial services sector, where these businesses have split up their central e-commerce units to respective business units after achieving a critical mass of expertise and associated policies and practices (that is, having passed though the second stage of electronic service provision maturity as discussed in Chapter 2).[34]

The advantages of such an elevation and corresponding progress along the electronic services maturity model discussed previously are as described here—in particular, the elevation and centralization of electronic services planning in a way that eliminates the need for multiple units to tackle the problem in different, potentially inconsistent, ways. The potential disadvantages stem from the understandable organizational friction and cost that may result from changing roles and responsibilities. A loss of autonomy may result in an inability to modify practices in a way that meets local needs and conditions. However, the committee believes that the most effective way forward is through the sorts of changes proposed here. Discussions with experts from and observations of various industry sectors have buttressed this point of view.[35]

[34]Financial institutions, after moving through the second phase, have come to recognize the need for more centralized planning and architectural control, to ensure that all customer transactions and assets across all businesses and products can be recognized and to provide the customer with a more customer-centric view—and consistency across all delivery channels—branch, call center, Web bank. This has led to a tiering—there is a consumer bank business, but it includes cards, mortgages, brokerage, and so on—so that there is some central control and integration of all these businesses and products when presented to the customer.

[35]One senior telecommunications executive described a similar transition at his company and argues that organizational structure and culture were key to this conversion. Indeed a revolution occurred at his company to achieve the transition. He pointed out that the revolution was dramatic, costly, and ultimately very effective in cost, capability, and fundamentally changing the company as an enterprise. The revolutionary e-business activity was developed in parallel with the conventional IT organization and was anointed by the

Finding: The SSA's present direction diverges from the three-phase progression that large financial institutions have followed in successfully developing and launching electronic services.

Recommendation: In order to move to the second phase of electronic services maturity, the SSA should create a focal point responsible for developing and managing electronic information and service delivery—including components such as Web content, online transactions, user interfaces, research, database systems and other key enabling technologies, and other facets of electronic service delivery that are currently dispersed throughout the SSA. This focal point should have sufficient resources to take on organization-wide responsibility for online services and should report directly to the SSA Commissioner or to a Deputy Commissioner.

Embracing Change

Although charting a roadmap for the future under these circumstances may be daunting, the terrain is not completely unexplored and uncharted. The SSA is a unique enterprise in some important ways, but as discussed in Chapter 2, there are many other ways in which its constraints, contexts, and clientele are similar to those under which other large-scale financial institutions must operate. The extensive stock of experiences and lessons learned by these other institutions should serve as a useful guide to the approaches that the SSA might take during its evolution toward electronic services. Studying these experiences leads the committee to a sharpened appreciation of the need to embrace and engage the overriding issue of change. However, the committee views any individual recommendation of a specific path toward a specific suite of services as subordinate to what the committee sees as a more general and overarching need: the SSA should embrace change as a factor in the way that it makes plans for the future and in the way that it does business.

Particularly as change in the various domains of SSA involvement continues, it will become increasingly important for the SSA to devote attention and resources to being well informed about the nature and ramifications of these changes, both for itself and for its various user

---

chief executive officer as the new IT head. The key to the success of the e-business activity was an e-business vision and a corresponding architecture designed to Web-enable as many services as possible. The vision and the architecture were required as a context for the many component technology solutions, but these would not have succeeded without the revolution. Personal communication from Michael Brodie, Distinguished Architect, Verizon Communications, to Lynette Millett, August 18, 2006.

groups, including beneficiaries. Although the SSA does pay attention to change and seeks advice externally and internally, there are opportunities to use that advice more systematically as a basis for sustained and effective action. Broader and more systematic attention to anticipating and addressing change is warranted.

Finding: The SSA faces significant ongoing change—in terms of technology, demographics, and public expectations—as it carries out its activities, services, and interactions with a variety of user communities.

Recommendation: The SSA should embrace change as a constant. It should regularly evaluate emerging trends in such areas as technology (for example, database technologies) and business practices (for example, by learning from the experiences of financial institutions and moving toward the use of strategic partnerships for efficiency and effectiveness). It should also regularly evaluate the changing societal attitudes and expectations of its various user communities. The SSA should also institutionalize the formulation of strategies for addressing these trends.

The aim of this report is to present a forward-looking vision and set of recommendations for the SSA to keep in mind as it strives to achieve the kind of technological and cultural transformation needed to best reflect and reinforce the existing and admirable, deeply grounded institutional commitment to its beneficiaries. The next section provides a brief description of governmental transformation conceived broadly that lays out a forward-looking vision for the SSA as it seeks to bring about the kinds of change that will be needed in the years and decades ahead.

THE PROSPECT OF GOVERNMENTAL TRANSFORMATION

As the SSA develops and implements its electronic services strategy, what the agency is attempting to achieve is taking place within a more comprehensive shift to electronic service provision across the federal government. The visions of e-government discussed below describe the transformative effect of automating and reengineering processes that enable the government delivery of information and services using technology in several different ways. (See Box 4.1 for specific prospects for IT-enabled process transformation at the SSA.)

This transformation might also be thought of as a convergence—of user experience, data, or organization. For example, *a convergence of user experience* might be the ability of users to access all SSA program information and transactions easily through a Web portal. *Data*

BOX 4.1
Information Technology-Enabled Process Transformation

In many cases in both the public and private sectors, organizations implement technologies in ways that implicitly reinforce the political status quo.[1] This phenomenon is referred to as "technology enactment." The term often (but not always) refers to the tendency of actors to implement new information technology (IT) in ways that reproduce or even strengthen institutionalized, sociostructural mechanisms. This can occur even when such enactments lead to seemingly irrational and suboptimal uses of technology.[2]

As a simple example, navigational principles for an institution's Web site may appear mysterious to an external user because the organization of information on the Web site mirrors the organizational chart of the institution rather than reflecting likely functionalities or information resources that external users might be interested in.[3,4] This suggests that if a Web site merely mirrors an organization's internal structure, if that organizational structure is not optimally designed for electronic services, the Web site will not be either. Similarly, most telephone book listings for government offices are organized by agency structure, not by citizens' needs. If one wants to complain about the speed of traffic in front of one's home, it is often not clear which city organization to call. Many of the calls to a general city number are in service of finding out just *whom* to call. A telephone book listing that met the user's goal (for example, road complaints) would eliminate the need for people whose primary job is to route calls.

A 1990 National Research Council (NRC) study committee made a number of recommendations related to Social Security Administration (SSA) systems mod-

[1] Jane Fountain, *Building the Virtual State: Information Technology and Institutional Change*, Washington, D.C.: Brookings Institution Press, 2001.

[2] Jane Fountain, *Building the Virtual State: Information Technology and Institutional Change*, 2001.

[3] Jane Fountain, *Building the Virtual State: Information Technology and Institutional Change*, 2001.

[4] At one of this study's open sessions, Ben Shneiderman, University of Maryland's Human Computer Interaction Laboratory, presented principles of good Web site design and methods for doing usability testing. He found the SSA Web site to be difficult to navigate intuitively. As discussed previously, a potential reason that the SSA Web site presents navigation challenges is that it may be mirroring the internal organizational structure of the SSA.

convergence might manifest itself in a user's being able to see all of his or her data and program eligibility information from the SSA on a single page within the portal. *Organizational convergence* would mean that users could understand the interrelationships among the various SSA programs from a user's perspective, not based on the SSA's organizational

ernization.[5] Some of those earlier recommendations can be applied to the SSA today in the context of rethinking the agency's strategy for service delivery. For example, that NRC committee found that a particular target of opportunity was disability processing, which in 1990 was almost exclusively a paper-based file operation that consumed a disproportionate share of operating budget. Although the SSA tried to redesign the disability processes during the 1990s, the implementation largely failed.[6] Today, while the agency is deploying an electronic disability system, that system largely automates the existing paper-based file processes, not capitalizing on the simplicity that the digital medium offers. The result of this approach is that users and the government get fewer benefits, such as reduced cycle times, lower error rates, and decreased cost per transaction, than might otherwise be possible through a more thorough transformation. The present committee does recognize that in early 2006 the SSA Commissioner announced a major overhaul and streamlining of the disability-determination processes.[7] The changes are largely process changes, and the extent to which strategy and planning for procedural and technical changes (for example, for the current Electronic Disability system, eDIB) were coordinated within the SSA is not clear.

In general, the SSA's various services are furnished through the execution of a variety of intricate processes. Other industries and financial institutions report impressive gains from disciplined overhauls of existing basic business processes, carried out in order to take advantage of new efficiencies and functionalities offered by information technology. Given its imminent workload and workforce challenges, the SSA would be well served by being proactive in transforming its processes to make the best use of both its human capital and its IT investments. Process transformation of this sort will of necessity be closely tied to and dependent on the modernization and overhaul of the database systems as described in Chapter 3 of this report.

---

[5]National Research Council, *Systems Modernization and the Strategic Plans of the Social Security Administration,* Board on Telecommunications and Computer Applications, Washington, D.C.: National Academy Press, 1990.

[6]See Social Security Advisory Board, *How SSA's Disability Programs Can Be Improved,* August 1998, pp. 6-8, available at http://www.ssab.gov/Publications/Disability/report6.pdf, accessed July 11, 2007; and Social Security Advisory Board, *Charting the Future of Social Security's Disability Programs: The Need for Fundamental Change,* January 2001, pp. 15-18, available at http://www.ssab.gov/Publications/Disability/disabilitywhitepap.pdf, accessed July 11, 2007.

[7]Social Security Administration, "News Release: Commissioner Barnhart Unveils New Social Security Disability Determination Process," March 28, 2006. For more information on the changes, see http://www.ssa.gov/disability-new-approach/.

structure. These types of convergence could bring to fruition the best possibilities of e-government. Depending on the underlying strategic goals of e-government, which may vary depending on, for example, the political leadership driving the changes, the attendant benefits of such convergence can take the form of streamlined government organizations, improved service delivery, and/or more participatory decision making.[36]

Convergence of User Experience and Data

The envisioned transformation of government information and service delivery begins with a radically transformed user experience. Web portals are one technology platform that is helping to bring a more user-centric perspective to government.[37] E-government offerings will increasingly target market segments (for example, students); life events (for example, buying a house); or role (for example, small business owner versus nonprofit treasurer) to help insulate users from the complexity of government programs and organizational structures. This transformation would require government to think "outside in" in terms of users needs (see Chapter 1 for examples) instead of focusing on how agencies are divided by program, level of government, or unit within level of government.

There is little more frustrating for an individual using government-provided electronic services than having to reenter the same data sets for different offices in one federal agency to complete one transaction. Privacy concerns notwithstanding, users may reasonably ask, "Why doesn't my government have its act together?" Users are becoming increasingly accustomed to and comfortable with private-sector organizations with which they transact business having a more holistic understanding of who they are as a customer. Whether it is a financial services organization or a retailer, users have come to expect that those businesses understand them as a customer across organizational units, products lines, or service channels. An existing customer of a large financial institution might reasonably expect that if one has car insurance and wants to buy homeowner's insurance from the same institution, a person will not be starting from scratch in filling out the application for the new product—at a minimum, an agent might provide a pre-filled-out application as a starting point. What enables this customer-centric view is data convergence. These businesses have either organized and consolidated their data centrally, or they have

[36]D.M. West, *Digital Government: Technology and Public Sector Performance*, Princeton, N.J.: Princeton University Press, 2005.

[37]See D.B. Gant, J.P. Gant, and C.L. Johnson, "State Web Portals: Delivering and Financing E-Service," Endowment for the Business of Government, Arlington, Va., Pricewaterhouse Coopers, January 2002, available at http://www.businessofgovernment.org/pdfs/JohnsonReport.pdf; and V. Jupp and S. Shine, "Government Portals—The Next Generation of Government Online," pp. 217-223 in *Proceedings of the First European Conference on E-Government*, Dublin, Ireland: Trinity College, 2003.

linked disparate data stores to create a holistic view of the customer, often relying on customer relationship management (CRM) systems.

Organizational Transformation

Although many large organizations have top-down decision making, limited front-line discretion, and specialized tasks, these characteristics tend to be more pronounced in government organizations. Government organizations typically reflect the Weberian ideals of specialization of task, limited front-line employee discretion, and top-down decision making.[38] These organizational and management principles reflect the prevailing wisdom when these institutions were reformed or created during the Industrial Revolution. What this means in practice, though, is that users may have to navigate through a bureaucratic maze to complete relatively straightforward transactions.

Consider an entrepreneur who would like to start a business. Such a person likely would have to register his or her business, obtain tax identification numbers, comply with regulatory requirements, and generally fill out paperwork for multiple agencies at the federal, state, and local levels. Ideally, the responsible agencies for regulating, taxing, and supporting small business generation would have converged to provide a "starter-kit" for new businesses that consolidated and rationalized paperwork, regulatory oversight, and reporting requirements. This would represent not only organizational transformation, but the data convergence and transformed user experience described earlier.

These transformations and the related level of convergence typically evolve over time. Although it may be possible for a government organization to outsource its electronic service capabilities and achieve order-of-magnitude improvements in convergence, such change more typically occurs incrementally. Several models have been developed to describe or predict such an evolution of e-government capabilities.[39] (Figure 4.1 presents as an example a four-stage model of e-government.) These models suggest that e-government capabilities begin modestly and initially provide static, one-way information, but grow more sophisticated and add interactive and transactional capabilities. The models predict the most

---

[38]Max Weber, "Bureaucracy," pp. 23-29 in *Classics of Public Administration*, Jay M. Shafritz and Albert C. Hyde, Eds., Oak Park, Ill.: Moore Publishing, 1978.

[39]See, for example, C. Baum and A.D. Maio, "Gartner's Four Phases of E-Government Model," Research Note, Stamford, Conn.: Gartner, November 21, 2000; Janine Hiller and Francine Bélanger, "Privacy Strategies for Electronic Government," pp. 162-198 in *E-Government 2001*, Mark A. Abramson and Grady E. Means, Eds., Lanham, Md.: Rowman and Littlefield, 2001; and Karen Layne and Jungwoo Lee, "Developing Fully Functional E-Government: A Four Stage Model," *Government Information Quarterly* 18(2):122-136, 2001.

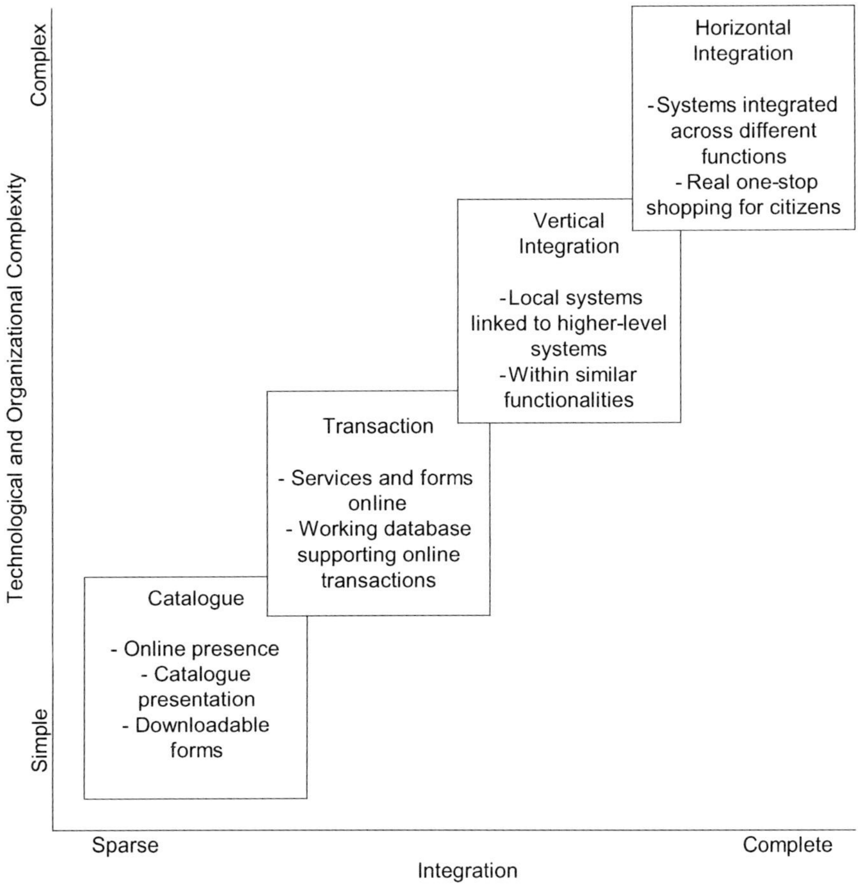

FIGURE 4.1 Layne and Lee's four-stage model of e-government.
SOURCE: Reprinted from Karen Layne and Jungwoo Lee, "Developing Fully Functional E-Government: A Four Stage Model," *Government Information Quarterly* 18(2):122-136, 2001, copyright 2001, with permission from Elsevier.

desired form of e-government, which includes horizontal and vertical integration of organization, data, and user experience that is fundamentally different from traditional forms of government service provision.

This kind of convergent, transformed public service is sometimes referred to as "joined up government"—a term coined in the 1990s in the United Kingdom.[40] Similar to the case in the three-stage model of

[40]See, for example, Vernon Bogdanor, *Joined-Up Government*, British Academy Occasional Papers, Oxford University Press/British Academy, Oxford, United Kingdom, July 21, 2005; and Andrea Di Maio, *What "Joined Up Government" Really Means*, Stamford, Conn.: Gartner, 2004.

transformation in the financial services industry, the higher the level of transformation and convergence, the higher the payoff for governments, taxpayers, and users. Joined up government also recognizes the roles of third parties in the delivery of government information and services, as described previously. As is the case with electronic commerce in the private sector, sometimes the government agency is only one part of a larger virtual value chain in which, for example, a taxpayer relies on a tax preparer to prepare and file taxes, or a veteran relies on a veterans' service organization to file a disability claim with the Department of Veterans Affairs.[41]

From "E-Government" to "Government"

Technology trends and the emerging areas of convergence described above all suggest that "e-government" may soon seem an old-fashioned term. The "e-government" or "electronic government"[42] labels are yesterday's news, because virtually all governmental and democratic processes are being or will be transformed by the strategic use of information and communication technologies. Perhaps the "e" preface can be thought of just as easily to signify "effective" or "efficient" government moving beyond "electronic." The committee noted that the concepts of "e-business" and "e-commerce" are no longer routinely used; in fact, these concepts are more routinely known as "business" and "commerce." So, too, it would seem that over the course of the next several years "e-government" could revert to "government," albeit "transformed" government or perhaps even "networked" government.

One could speculate that many government agencies, including the SSA, have let the concept of "e-government" continue to deter them from undergoing a true transformation. Once an agency has done the (comparatively) easy part, that is, making information available through Internet and Web technologies, the agency in essence has provided citizens with access to the agency's information and services by way of the

[41]S.H. Holden and P.D. Fletcher, "The Virtual Value Chain and E-Government Partnership: Non-Monetary Agreements in the IRS E-File Program," pp. 375-394 in *Handbook of Public Information Systems*, G.D. Garson, Ed., New York: Marcel Dekker, 2005; and K. Frey and S.H. Holden, "Distribution Channel Management in E-Government: Addressing Federal Information Policy Issues," *Government Information Quarterly* 22(4):685-701, 2006.

[42]"E-government" or "electronic government" has many working definitions. For example, Sharon Dawes, director of the Center for Technology in Government at Albany, defines e-government as "the use of information technology to support government operations, engage citizens, and provide government services." And the European Union defines e-government as "the use of information and communication technology in public administrations combined with organizational change and new skills in order to improve public services and democratic processes and strengthen support to public policies."

"front office." The challenging part is for agencies to strategically transform "back-office" operations by using information and communication technologies to undertake the more complex processes of reengineering the core business processes, associated policies, and ultimately the services themselves; channel integration and management that successfully merges the various ways that citizens will want to access services will be essential. Much work over the past few years has explored the implications of this needed strategic transformation. One public administration expert states:

> Many assume e-government is solely about delivering government services over the Internet. This popular assumption is very limited for two reasons. First, it narrows our vision for e-government because it does not allow for the wide range of governmental activities that are not direct services; nor does it recognize the essential use of technologies other than the Internet. Second, it grossly oversimplifies the nature of e-government, leaving the impression that a nicely designed, user-oriented web site is the whole story. This ignores the substantial investments that are needed in people, tools, policies, and processes. It fails to recognize that while the citizen sees e-government from the public side of a web site or email screen, the real *work* of e-government is on the other side, inside the government itself.[43]

The committee has observed throughout this report that the SSA's e-government strategy focuses primarily on the "front-office" services and seems to be missing or too slowly tackling "back-office" transformation.

A Forrester research report published in 2005 finds that, looking toward the future, there are significant opportunities in e-government as agencies address obstacles such as persistent agency silos and the eroding career IT workforce within government. The report concludes that "over time, agencies will obtain sustainable results from e-government initiatives by adopting best practices like addressing process change before implementing technologies and providing integrated services through multi-agency portals."[44] Similarly, this committee concludes that the SSA's e-government strategy could be improved by reengineering processes and

---

[43]Sharon Dawes is director of the Center for Technology in Government at Albany (see http://www.ctg.albany.edu). Her article, "The Future of E-Government," is based on testimony presented to the New York City Council Select Committee on Information Technology in Government's hearing, "An Examination of New York City's E-Government Initiatives," June 24, 2002, available at http://www.ctg.albany.edu/publications/reports/future_of_egov, accessed July 11, 2007.

[44]See Alan E. Webber, with Bradford J. Holmes, Gene Leganza, and Sara E, McAulay, "The Future of E-Government: Introducing the e-Government Maturity Continuum," available at http://www.forrester.com/go?docid=36950, accessed July 11, 2007.

close collaboration with other government agencies and third-party organizations when serving the public.[45]

One examination of challenges for e-government found that e-government agendas are merging with wider agendas about the role and scope of government. In the 1990s, three broad trends dominated much discussion of government: globalization, marketization, and technology. Their linked effects promised smaller government and increased customer choice. However in the 2000s these trends have blurred. As e-government initiatives move beyond tightly focused transactions, they confront issues of integration, information sharing, ethics, access, equity, and governance.[46]

Thus, it is increasingly clear that government transformation cannot happen in isolation, because true transformation requires working beyond the agency's walls, literally and figuratively. As Stephen Goldsmith, former mayor of Indianapolis, describes his vision of networked government, "It is not about outsourcing vs. bureaucracy it is about managing diverse webs of relationships to deliver value."[47] So too, have the Australians started thinking in terms of "networked governance": "While the public may traditionally have thought of government as synonymous with bureaucracy, in the future government will be highly networked and delivering outcomes through federations of organizations and agencies."[48]

---

[45]The OMB's Federal Enterprise Architecture envisions these transformations, as detailed in a 2004 OMB report on the Federal Enterprise Architecture (FEA) Program. (See Office of Management and Budget, *Expanding E-Government: Partnering for a Results Oriented Government*, December 2004; available at http://www.whitehouse.gov/omb/budintegration/expanding_egov12-2004.pdf, accessed July 11, 2007.) The FEA program builds a comprehensive business-driven blueprint of the entire federal government. The development of this framework has and will continue to encourage the federal government to identify opportunities to take advantage of technology to reduce redundancy; facilitate horizontal (cross-federal) and vertical (federal, state, and local) information sharing; establish a direct relationship between IT and mission/program performance to support citizen-centered, customer-focused government; and maximize IT investments to better achieve mission outcomes. The FEA framework and its five supporting reference models (Performance, Business, Service, Technical, and Data) are now used by departments and agencies in developing their budgets and setting strategic goals.

[46]Robert Smith, "Centralization and Flexibility in Delivering E-Services: Tensions and Complements," Discussion Paper No. 19, available at http://www.agimo.gov.au/publications/2004/05/egovt_challenges/issues, accessed July 11, 2007.

[47]William D. Eggers and Stephen Goldsmith, "Networked Government," pp. 28-33 in *Government Executive*, June 2003, available at http://www.manhattan-institute.org/pdf/gov_exec_6-03.pdf, accessed July 11, 2007.

[48]John Halligan and Trevor Moore, "Future Challenges for E-Government," Monograph, available at http://www.agimo.gov.au/publications/2004/05/egovt_challenges/overview, accessed July 11, 2007.

CONCLUSION

The SSA faces formidable challenges in attempting to continue to provide its services predominantly through face-to-face personal interactions. In the years and decades ahead, its chief challenges will be the need to be prompt and reliable, to guarantee security and privacy in its transactions, to support the accurate execution of extremely large numbers of transactions, and to offer its services in ways that are convenient to the client, yet efficient. IT and electronic services are the more likely vehicles for meeting these challenges and for executing an effective campaign to provide an expanding portfolio of services to larger and increasingly diverse users. The SSA and its clients cannot afford to allow cultural impediments to keep the agency from embracing the changes that are so clearly needed and indicated.

Mechanisms such as monitoring and observing the societal context, keeping track of technology, establishing quantitative goals, putting in place effective measurement vehicles, and maintaining close, cordial, and continuous ties with all user communities support continuous change and process transformation within organizations. Although pursuing the creation of these mechanisms seems essential, a more urgent need is an appraisal of the SSA's readiness to pursue them. This report suggests an immediate examination of the agency's management structure, its decision-making processes, and its organizational culture, to ensure that all are poised to support the shift toward a culture that continuously strives to meet effectively the manifold pressures for change.

Although the challenges outlined in this report are numerous and sizable, the committee is confident that they are not insurmountable for the SSA. The committee's investigations made it clear that the SSA and its people are firmly dedicated to meeting beneficiaries' needs with enthusiasm and professionalism. Their dedication to their mission seems absolute and unwavering. This report is offered in the spirit of advice to dedicated professionals regarding opportunities to do ever better in meeting continually growing challenges.

Appendixes

A

Committee and Staff Biographies

COMMITTEE MEMBERS

LEON J. OSTERWEIL, *Chair,* is a professor in the Department of Computer Science, University of Massachusetts, Amherst. He is a fellow of the Association for Computing Machinery (ACM), has been an ACM lecturer, and has served on the editorial boards of *IEEE Software* and ACM's *Transactions on Software Engineering and Methodology*. He was interim dean of the College of Natural Sciences and Mathematics at the University of Massachusetts, Amherst, where he is the codirector and cofounder of both the Electronic Enterprise Institute and the Laboratory for Advanced Software Engineering Research. Dr. Osterweil was chair of the Computer Science Department at the University of Colorado at Boulder and of the Information and Computer Science Department of the University of California, Irvine, where he was also the founding director of the Irvine Research Unit in Software. He has been the program chair of many conferences, including the 16th International Conference on Software Engineering. He was General Chair of both the 28th International Conference on Software Engineering and the Sixth International Symposium on the Foundations of Software Engineering. Dr. Osterweil's research has centered on the effective application of computing to support software development, electronic commerce, medical safety, and digital government. His research has placed particular emphasis on process programming, tool and environment integration mechanisms, and software testing and analysis tools. He has been a principal investigator on a number of National Science Foundation (NSF) and Advanced Research Projects

Agency/Defense Advanced Research Projects Agency (ARPA/DARPA) projects over the past 35 years. He has published and presented more than 100 papers on a variety of software engineering topics, mostly in leading venues worldwide. He has consulted for such organizations as IBM, Bell Laboratories, SAIC, MCC, KLA-Tencor, Tata Consultancy Services (India), The Numerical Algorithms Group, Ltd (UK), and TRW. He was a member of the Process Program Advisory Board of Carnegie Mellon University's Software Engineering Institute for 8 years. He received his A.B. from Princeton University and his M.S. and Ph.D. degrees from the University of Maryland, all in mathematics.

MATT BISHOP is a professor in the Department of Computer Science at the University of California, Davis, and a codirector of the Computer Security Laboratory there. His main research area is the analysis of vulnerabilities in computer systems, including modeling them, building tools to detect vulnerabilities, and ameliorating or eliminating them. He is active in information assurance education and is a charter member of the Colloquium on Information Systems Security Education. He has been active in the area of UNIX security since 1979 and has presented tutorials at SANS, USENIX, and other conferences. His textbook, *Computer Security: Art and Science*, was published in December 2002 by Addison-Wesley Professional. Dr. Bishop also teaches software engineering, machine architecture, operating systems, programming, and computer security. In 2006, he received the Colloquium Academia Award for his research and pedagogical work in computer information security. He received his Ph.D. in computer science from Purdue University, where he specialized in computer security, in 1984.

MICHAEL J. CAREY received his Ph.D. in computer science from the University of California, Berkeley, in 1983. From 1983 to 1995 he was on the Computer Sciences Department faculty at the University of Wisconsin-Madison, where he taught and conducted research on a variety of database system architecture and performance issues. In 1995, Dr. Carey moved to IBM Research, where he worked on data integration, DB2 Universal Database, and Extensible Markup Language (XML) query technologies and managed a small team of object/relational and XML database researchers. Dr. Carey left IBM in 2000 to lead an e-commerce infrastructure team as a fellow at Propel, a small, Internet-era startup company. He joined BEA Systems, Inc., in late 2001, and since that time he has been working on the XML data-handling and XQuery-related aspects of BEA's WebLogic Integration, Liquid Data for WebLogic, and AquaLogic Data Services Platform products. He is currently a senior engineering director and product architect within BEA's AquaLogic product organization. Dr. Carey has authored well over 100 technical articles in refereed conference proceedings and journals. He has served on the editorial boards of ACM's *Transactions on Database Systems*, *VLDB Journal*, and *IEEE Database Engineering Bulletin*, and he has served on numerous conference committees. He is co-editor-in-chief of Springer-Verlag's book series on Data-Centric Systems and Applications and is currently serving on the ACM SIGMOD (Special Interest Group on Management of Data) Awards Committee and the University of Wisconsin (UW)-Madison Computer Sciences Department Board of Visitors. Dr. Carey has received a number of awards, including an IBM Faculty Development Award, a Digital Equipment Corporation Incentives for Excellence Award, a UW-Madison Computer Sciences Department Teaching Award, an NSF Presidential Young Investigator Award, an IBM Outstanding Technical Achievement Award, a VLDB (Very Large Databases) 10-Year Best Paper Award, an ACM SIGMOD Test of Time Paper Award, a University of California, Berkeley, Computer Science Outstanding Alumnus Award, and the ACM SIGMOD Contributions and Edgar F. Codd Innovations Awards. Dr. Carey is a fellow of the ACM and a member of the National Academy of Engineering.

DAVID J. DeWITT joined the Computer Sciences Department at the University of Wisconsin-Madison in September 1976 after receiving his Ph.D. degree from the University of Michigan. He served as department chair for 5 years (July 1999 to July 2004) and is currently the John P. Morgridge Professor of Computer Sciences. In 1995, Professor DeWitt was selected to be an ACM Fellow. He also received the 1995 SIGMOD Innovations Award for his contributions to the database systems field. He was elected to the National Academy of Engineering in 1998. His research program has focused on the design and implementation of database management systems including parallel, object-oriented, and object-relational database systems. In the late 1980s, his Gamma parallel database system project produced many of the key pieces of technology that form the basis for today's generation of large parallel database systems, including products from IBM, NCR/Teradata, Netezza, Tandem, and Oracle. Throughout his career, he has also been interested in database system performance evaluation. In the early 1980s, he developed the first relational database system benchmark, which became known as the Wisconsin benchmark. More recently, his research program has focused on the design and implementation of distributed database techniques for executing complex queries against the content of the Internet. Professor DeWitt has authored more than 120 technical publications and served on numerous program committees and NSF review panels. He was a member of the NSF's Computer and Information Science and Engineering (CISE) Advisory Committee (2000-2003) and has served on several National Research Council (NRC) and Defense Advanced Research Projects Agency (DARPA) study panels. He is currently a member of the Computer Science and Telecommunications Board (CSTB). He was the program chair of the 1983 SIGMOD conference, program co-chair of the 1988 VLDB conference, and general chair of the 2002 SIGMOD conference. He has graduated 32 Ph.D. students. Professor DeWitt has served as a consultant to numerous companies, including IBM, NCR, Informix, Tandem, and Microsoft, on a variety of technical issues regarding the design and implementation of database systems.

VALERIE GREGG is assistant director for development at the Digital Government Research Center of the Information Sciences Institute at the University of Southern California. She is secretary of the North American chapter of the Digital Government Society. She is co-principal investigator on a National Science Foundation-funded 4-year award entitled "Building and Sustaining an International Digital Government Research Community of Practice." Prior to working in academia, Ms. Gregg had a 30-year career in public service at the federal level. For 8 years, she was program manager for the Digital Government Research program in the Division of Information and Intelligent Systems at NSF. She has been on the conference committee for the annual International Conference for Digital Government Research since its inception in 2000. Prior to her service at NSF, Ms. Gregg worked for 22 years at the U.S. Census Bureau as a program manager in various aspects of the decennial census. She was a member of the Joint Ventures Project with the Census Bureau, and her work included involvement with the Tiger Mapping Service, a coast-to-coast digital map base that was designed and implemented to demonstrate cost-efficient delivery of public data and research and development of Census Bureau applications on the Internet. She also served as chair for the Interagency Task Force responsible for the design, development, and management of the award-winning Internet site www.fedstats.gov for "one-stop shopping" for federal statistics.

BLAISE HELTAI is a partner at NewVantage Partners focusing on the applications of technology to business transformation, emerging technologies, multichannel strategy, and Internet delivery. Dr. Heltai was previously an executive at the FleetBoston Financial Group, where, as the managing director for e-business, he led the company through a fundamental transformation of e-enablement, integration, and rationalization of Internet initiatives for all lines of business—consumer, commercial, and government—in the United States and internationally. He had also led customer relationship management (CRM) initiatives for FleetBoston, as well as managing customer data analysis, direct marketing, market research and research and development (R&D) functions for Fleet and Bank of America. He was chief executive officer of fileTRUST, which produced a secure online document-storage and -sharing service. He has held positions at Bell Laboratories and at AT&T Consumer Products, where he led efforts ranging from assessing the demand for new telecommunications services to inventing, developing, and managing interactive television and online services. Dr. Heltai has been published and widely quoted by the national print and broadcast media as a recognized expert in areas such as e-business, online banking and investing, public policy issues, payment technologies, video on demand, and emerging technologies. He has won numerous awards, including Internet Week's Top 100 of 2001, MassEcomm Top 10 Executives, Microsoft Innovation Award, and Best in Show at the Consumer Electronics Show. He has served on the board of directors of MECA Software, Integrion, LLC, and numerous nonprofit organizations. He has been on the advisory boards of S1 Corporation, PostX, and FTVentures, and currently serves as the president of the Massachusetts Innovation and Technology Exchange. He received his Ph.D. in mathematics from the State University of New York at Stony Brook in 1984.

STEPHEN H. HOLDEN is a principal with Touchstone Consulting Group. While participating on the present NRC committee, he was an assistant professor in the information systems department at the University of Maryland, Baltimore County (UMBC). Dr. Holden's research included the management of information technology in the public sector, information policy, electronic authentication policies and practices, and electronic government. He has published in *Administration and Society, IEEE Internet Computing, Government Information Quarterly, The Information Society, International Journal of Public Administration*, and *Public Performance and Management Review*. He has also contributed book chapters on information policy, electronic government, and information technology in the public sector. Dr. Holden was on the CSTB committee that produced the report *Who Goes There?: Authentication Through the Lens of Privacy*. He also serves on the editorial board of the *Journal of Public Administration Research and Theory* and the *International Journal of E-Government Research*. The MITRE Corporation, the IBM Center for the Business of Government, and the Real User Corporation have funded his research. Prior to his academic career, Dr. Holden worked for 16 years in the federal government. That experience included joining the civil service as a presidential management intern at the Naval Sea Systems Command. He spent 10 years at the Office of Management and Budget doing a variety of policy and management work. In his last federal position, he worked at the Internal Revenue Service as a member of the senior executive service in the Electronic Tax

Administration organization. He earned his Ph.D. in public administration and public affairs from Virginia Polytechnic Institute and State University. He also holds a master of public administration and a bachelor of arts in public management from the University of Maine.

LARRY G. MASSANARI was acting commissioner of Social Security and most recently served as the regional commissioner for the Social Security Administration's (SSA's) Philadelphia region, where he directed agency operations in the mid-Atlantic states. Mr. Massanari began his career with Social Security in 1966 as a claims representative in one of SSA's field offices. He served in a variety of administrative and managerial positions in the Chicago area before moving to the agency's headquarters in 1979. He held several top-level positions in SSA's central office, including director of human resources and associate commissioner for policy. He was named regional commissioner for the Philadelphia Region in 1987. Mr. Massanari has received numerous awards for outstanding achievement. He was a Presidential Executive–rank award winner in 1992 and 1997, receiving the rank of both Meritorious and Distinguished Executive. He was also named a finalist by the Professional Development League in 1997 for distinguished executive service. He has been the recipient of three Commissioner's Citations, the highest honor conferred on Social Security Administration employees. In 1990, he received the first Commissioner's Leadership Award, which recognizes the accomplishments of SSA executives. A graduate of Western Michigan University, Mr. Massanari also completed graduate work at Northwestern University.

JUDITH S. OLSON is the Richard W. Pew Chair in Human Computer Interaction at the University of Michigan, Ann Arbor. She is also a professor in the School of Information, the Business School, and the Department of Psychology. Her research interests include computer-supported cooperative work, human-computer interaction, the design of business information systems for organizational effectiveness, and cognitive psychology. Dr. Olson's recent research focuses on the nature of group work and the design and evaluation of technology to support it. This field combines cognitive and social psychology with the design of information systems. Dr. Olson began her career at the University of Michigan in the Department of Psychology, served as a technical supervisor for human factors in systems engineering at Bell Laboratories in New Jersey, and returned to Michigan to the Business School and the then-new School of Information. She has more than 60 publications in journals and books, and served on the NRC committee that produced the report *Who Goes There?: Authentication Through the Lens of Privacy*. She has recently been appointed to the CHI Academy of ACM's Special Interest Group on Computer-Human

Interaction. Dr. Olson earned a B.A. in mathematics and psychology from Northwestern University in 1965 and a Ph.D. in 1969 in the same disciplines from the University of Michigan.

STAFF

LYNETTE I. MILLETT is a senior program officer and study director at the Computer Science and Telecommunications Board (CSTB) of the National Research Council. She is currently involved in several CSTB projects, including a study on sustaining growth in computing performance, an exploration of software-intensive systems producibility, and an activity on biometrics systems, among other things. She was the study director for the CSTB projects that produced *Software for Dependable Systems: Sufficient Evidence?*, *Who Goes There?: Authentication Through the Lens of Privacy*, and *IDs—Not That Easy: Questions About Nationwide Identity Systems*. Her portfolio includes significant portions of CSTB's recent work on software and on identity systems and privacy. She has an M.Sc. in computer science from Cornell University and a B.A. in mathematics and computer science with honors from Colby College. Her graduate work was supported by both a National Science Foundation graduate fellowship and an Intel graduate fellowship.

JOAN D. WINSTON is a program officer at the Computer Science and Telecommunications Board (CSTB) of the National Research Council. Before joining the staff of CSTB, she was an assistant director (Information Technology Team) at the U.S. Government Accountability Office. From 1998 to 2001, Ms. Winston was principal associate at Steve Walker and Associates, LLC, which managed early-stage venture funds focusing on information technology. From 1995 to 1998, she was director of policy analysis for Trusted Information Systems, Inc. From 1986 to 1995, she held various analytical and project direction positions at the congressional Office of Technology Assessment (OTA) and was named an OTA senior associate in 1993. Before her service with OTA, she worked briefly for the Congressional Research Service of the Library of Congress. Ms. Winston started her career as an engineer at the Charles Stark Draper Laboratory, Inc., in Cambridge, Mass. She received an S.B. in physics and an S.M. in technology and policy, both from the Massachusetts Institute of Technology.

JANICE M. SABUDA is a senior program assistant at the Computer Science and Telecommunications Board of the National Research Council. She has supported board activities and is involved in several studies, including *Improving Cybersecurity Research in the United States* and *Privacy* *in the Information Age*. Previously, she focused on the congressionally requested study that resulted in *Youth, Pornography, and the Internet* and the project that resulted in *Global Networks and Local Values: A Comparative Look at Germany and the United States*. Prior to joining CSTB in August 2001, she worked as a customer service representative at an online fund-raising company and as a client services analyst at a prospect research firm. She is currently pursuing a certificate in event management from the George Washington University Center for Professional Development. She received her bachelor of science degree (1999) in business administration from the State University of New York College at Fredonia.

B

Panelists and Briefers at Open Committee Meetings

OCTOBER 11-12, 2005, AND FEBRUARY 8-9, 2006

Panelists and Briefers

Diana Andrews, Social Security Administration
Jo Armstrong, Social Security Administration
Philip Becker, Social Security Administration
Jerry Berson, Social Security Administration
Mark Blatchford, Social Security Administration
Darlynda Bogle, Social Security Administration
Joseph Coates, Consulting Futurist
Jim Courtney, Social Security Administration
Renato J. DiPentima, SRA International
Marti Eckert, Social Security Administration
Debby Ellis, Social Security Administration
Marty Ford, Disability Policy Coalition
Judy Fryback, Disability Determination Service, Wisconsin
Lyman Goon, Social Security Administration
William E. Gray, Social Security Administration
Leonard Green, Rhode Island Department of Health
Myrtle S. Habersham, Social Security Administration
John Halamka, Harvard Medical School
Blaise Heltai, New Vantage Partners
John Horrigan, Pew Internet and American Life Project
Thomas P. Hughes, Social Security Administration

Steve Kautsch, Social Security Administration
Marcia Kent, Social Security Administration
Amy Lee, AARP Services, Inc.
Wayne Lemon, Social Security Administration
James B. Lockhart III, Social Security Administration
Terry Lutes, IRS (retired)
Keith Mallinson, Yankee Group
Denise Maynard, Social Security Administration
Linda S. McMahon, Social Security Administration
Paul Patrick, BEA Systems
Martin Prahl, Lockheed Martin
Peter Quinn, Former CIO, Commonwealth of Massachusetts
Ann Robert, Disability Determination Service, Illinois
Sue Roecker, Social Security Administration
Robin Sabatino, Social Security Administration
John Sabo, Computer Associates
Ben Shneiderman, University of Maryland, College Park
Dale W. Sopper , Social Security Administration
Terry Stradtman, Social Security Administration
Fritz Streckewald, Social Security Administration
Thomas Sutton, Immediate Past President, National Organization of Social Security Claimants' Representatives
Tony Tullo, ADP Tax and Financial Services
Nancy Veillon, Social Security Administration
Don Watson, Social Security Administration
Reginald F. Wells, Social Security Administration
Bill Zielinski, Social Security Administration

Social Security Administration Staff Providing Information, Material, and Assistance

Sean Brune, Social Security Administration
Bruce W. Carter, Social Security Administration
Gerard R. Hart, Social Security Administration
Renee Trujillo, Social Security Administration

C

Social Security Administration Major Office Missions

This information is quoted from material available on the Social Security Administration (SSA) public Web site.[1]

OFFICE OF THE COMMISSIONER OF SOCIAL SECURITY

The Office of the Commissioner (OC) is directly responsible for all programs administered by SSA; for State-administered programs directed by SSA; and for certain functions with respect to the black lung benefits program. It provides executive leadership to SSA. The Office is responsible for development of policy, administrative and program direction, program interpretation and evaluation, maintenance of relations with news media, research oriented to the study of the problems of economic insecurity in American society; and development of recommendations on methods of advancing social and economic security through social insurance and related programs.

OFFICE OF THE DEPUTY COMMISSIONER, COMMUNICATIONS

The Office of the Deputy Commissioner, Communications (ODCComm) is the SSA component responsible for the conduct of the Agency's national public information/public affairs (PI/PA) programs. Performs SSA

[1]The material in this appendix is quoted from mission statements in "Organizational Structure of the Social Security Administration," http://www.ssa.gov/org/ssaorg.htm, accessed May 30, 2007.

Press Office function to ensure a unified and consistent message to SSA's many publics. Provides guidance and direction from a PI/PA standpoint to the development of Agency policies and decisions and assesses their potential impact on SSA's customers, stakeholders and employees. Creates, develops, facilitates, implements, oversees and evaluates all SSA communications and PI/PA activities, both internal and external. Cultivates and maintains effective working relationships with a wide range of national organizations, advocacy groups, other Federal agencies, State and local governments, the White House, and the media. Promotes full and open participation in the communications process between and among SSA's customers and stakeholders at all levels. Coordinates the non-English communications activities within SSA. Additionally, responds to high priority correspondence and public inquiries; maintains an evaluation program that measures efforts to meet the communications needs of SSA's customers, stakeholders and employees; produces PI/PA material designed to provide SSA's various audiences with timely information about Social Security programs, protections, rights and responsibilities and related issues; utilizes state-of-the-art media, methods and technology in product development and dissemination and fully supports headquarters and field employees who are directly or indirectly involved in SSA PI/PA activities nationwide.

OFFICE OF DISABILITY ADJUDICATION AND REVIEW

The Office of the Deputy Commissioner for Disability Adjudication and Review (DCDAR) administers the nationwide Disability Adjudication and Review program for SSA. Provides the basic mechanisms through which individuals and organizations dissatisfied with determinations affecting their rights to and amounts of benefits or their participation in programs under the Social Security Act may administratively appeal these determinations in accordance with the requirements of the Administrative Procedure and Social Security Acts. DCDAR includes a nationwide field organization staffed with Administrative Law Judges (ALJs) who conduct impartial hearings and make decisions on appeals filed by claimants, their representatives, providers-of-service institutions and others under the Social Security Act. The Appeals Council of DCDAR impartially reviews ALJ decisions, either on the Appeals Council's own motion or at the request of the claimant, and renders the Commissioner's final decision when review is taken. Reviews new court cases to determine whether the case should be defended on the record or the Commissioner should seek voluntary remand, and reviews final court decisions in light of the programmatic and administrative implications involved and makes recommendations as to whether appeal should be sought. Provides advice and recommendations on Social Security Administration program policy and related matters, including proposed Social Security Rulings.

OFFICE OF DISABILITY AND INCOME SECURITY PROGRAMS

The Deputy Commissioner, Disability and Income Security Programs is the principal advisor to the Commissioner of Social Security on program policy issues and is involved in strategic planning, policy development, and analysis of SSA program policy. The Office of the Deputy Commissioner, Disability and Income Security Programs (ODCDISP) directs the formulation of program policy for SSA. It directs and manages the planning, development, issuance, and evaluation of operational policies, standards, and instructions for the Retirement and Survivors Insurance, Disability Insurance, Supplemental Security Income (SSI) program, and other SSA programs. The Office manages SSA's disability and SSI policy and research agendas and long-term disability initiatives. The Office assists in achievement of consistency in program policy across programs administered by SSA. The Office is involved in analyses of legislative and regulatory specifications and budgetary impacts of legislation on programs administered by SSA. The Office produces data on the programs of the Agency. It is involved in the development of demonstrations and studies that provide recommendations on program improvements. The Office is responsible for the Agency's Regulatory Program. Develops and implements policies and procedures and coordinates activities relating to the operation of Social Security programs outside of the United States.

OFFICE OF THE DEPUTY COMMISSIONER, BUDGET, FINANCE AND MANAGEMENT

The Office of the Deputy Commissioner Budget, Finance and Management (ODCBFM) directs the administration of comprehensive SSA management programs including budget, acquisition and grants, facilities management and publications and logistics. The Office directs the development of Agency policies and procedures as well as the management of the Agency financial management systems.

OFFICE OF THE DEPUTY COMMISSIONER, HUMAN RESOURCES

The Office of the Deputy Commissioner, Human Resources (ODCHR) directs the administration of comprehensive SSA human resources programs including: human capital and planning initiatives, personnel management, labor management relations, employee relations, civil rights and equal opportunity, and training.

OFFICE OF THE DEPUTY COMMISSIONER, LEGISLATION AND CONGRESSIONAL AFFAIRS

The Office of the Deputy Commissioner, Legislation and Congressional Affairs develops and conducts the legislative program of SSA, serves as the focal point for all legislative activity in SSA, analyzes legislative and regulatory initiatives and develops specific positions and amendments. The Office evaluates the effectiveness of programs administered by SSA in terms of legislative needs, and analyzes and develops recommendations on related income maintenance, social service and rehabilitation program proposals, particularly those which may involve coordination with SSA-administered programs, and on other methods of providing economic security. It provides advisory service to SSA officials on legislation of interest to SSA pending in Congress. It also provides legislative drafting to officials within the Executive Branch, congressional committees, individual Members of Congress and private organizations interested in Social Security legislation. It establishes and maintains a working relationship with all Members of Congress. It serves as SSA's information gathering and dissemination staff on congressional activities affecting SSA programs and handles certain claims and administrative matters that are particularly urgent or sensitive to Members of Congress.

OFFICE OF THE DEPUTY COMMISSIONER, OPERATIONS

The Office of the Deputy Commissioner, Operations (ODCO) directs and manages central office and geographically dispersed operations installations. It oversees regional operating program, technical, assessment and program management activities. It directs studies and actions to improve the operational effectiveness and efficiency of its components. It promotes systems and operational integration and defines user needs in the strategic planning process. It determines automation support needs for Operations components. It oversees the coordination and implementation of SSA's policies for the electronic delivery of Agency services to the public. This Office defines user concerns in the development of operational and programmatic specifications for new and modified systems, including the evaluation and implementation phases. When mutually agreed, provides support to the Office of Disability Adjudication and Review (ODAR) and/or specific State Disability Determination Services. Provides budget and management guidance for the disability claims activities as carried out by the State Disability Determination Services (DDS).

OFFICE OF THE DEPUTY COMMISSIONER, POLICY

The Deputy Commissioner, Policy is the principal advisor to the Commissioners of Social Security on major policy issues and is responsible for major activities in the areas of strategic policy planning, policy research and evaluation, statistical programs, and overall policy development and analysis. The Office of the Deputy Commissioner, Policy (ODCP) serves as the Agency lead spokesperson in presenting policy proposals and analysis within and outside the Executive Branch. The Office directs the formulation of overall policy for SSA and ensures the consistency of policy development and implementation activities across all programs administered by SSA. The Office broadly formulates, promulgates and interprets programs, objectives, and policy. The Office directs research, evaluation and analysis, and development of demonstrations and studies supporting the policy development of SSA; provides recommendations on modification of social insurance and income assistance programs administered by SSA; and conducts the statistical programs of the Agency. The Office is involved in developing legislative and regulatory specifications and analyses of legislative and budgetary impacts. The Office works with the Department of Treasury on issues of policy relating to the Federal Insurance Contributions Act and the Self-Employment Contributions Act, including such matters as definition of wages and implementation of laws. It directs formulation of Agency policy regarding related government programs that affect SSA programs and/or operations and negotiates related agreements with other agencies. It evaluates the effectiveness of national policies in meeting both short and long-term program goals.

OFFICE OF THE DEPUTY COMMISSIONER, SYSTEMS

The Office of the Deputy Commissioner, Systems (ODCS) directs the conduct of systems and operational integration and strategic planning processes, and the implementation of a comprehensive systems configuration management, data base management and data administration program. Initiates software and hardware acquisition for SSA and oversees software and hardware acquisition procedures, policies and activities. Directs the development of operational and programmatic specifications for new and modified systems, and oversees development, validation and implementation phases.

OFFICE OF THE CHIEF ACTUARY

The Office of the Chief Actuary (OCACT) plans and directs a program of actuarial estimates and analyses pertaining to the SSA-administered retirement, survivors and disability insurance programs and supplemental security income program and to projected changes in these programs. Evaluates operations of the Federal Old-Age and Survivors Insurance Trust Fund and the Federal Disability Insurance Trust Fund; estimates future operations of the trust funds; conducts studies of program financing; performs actuarial and demographic research on social insurance and related program issues; and estimates future workloads. Provides technical and consultative services to the Commissioner, the Board of Trustees of those two Trust Funds, and, as requested, congressional committees. Appears before congressional committees to provide expert testimony on the actuarial aspects of Social Security issues.

OFFICE OF THE CHIEF INFORMATION OFFICER

The Office of the Chief Information Officer (OCIO) develops the Information Resource Management Plan and defines the Information Technology (IT) vision and strategy for the Social Security Administration. The Office shapes the application of technology in support of the Agency's Strategic Plan including the Information Technology Architecture that outlines the long term Strategic Architecture and Systems Plans for the Agency and includes Agency IT Capital Planning. The OCIO supports and manages pre and post implementation reviews of major IT programs and projects as well as project tracking at critical review points. The OCIO provides oversight of major IT acquisitions to ensure they are consistent with Agency architecture and with the IT budget, and is responsible for the development of Agency IT security policies. The Office directs the realization of the Agency's Information Technology Architecture to guarantee architecture integration, design consistency, and compliance with federal standards, works with other Agencies on government-wide projects such as e-GOVERNMENT, and develops long range planning for IT Human Resource Strategies.

OFFICE OF THE CHIEF STRATEGIC OFFICER

The Office of the Chief Strategic Officer directs the administration of SSA's comprehensive management programs including strategic planning, workforce analysis, and competitive sourcing. It directs the development of the Agency's tactical and strategic planning process, the Agency Strategic Plan, Annual Performance Plan and Annual Performance Report; and tracks Agency performance in relation to established performance measures. Working with all components, the office identifies those priority initiatives needed to meet agency goals, objectives and outcomes and how to link these to budget input so that they can be funded and the outcomes achieved. It directs, develops and implements a comprehensive program of management studies, research and analyses. This allows SSA to evaluate and determine the feasibility of implementing major changes affecting the SSA organization, its administrative practices, its methods of operation and work processes and procedures, workflow and workload processing positions. It directs the Agency's policies and procedure as well as the management of the Agency competitive sourcing program.

OFFICE OF THE GENERAL COUNSEL

The Office of the General Counsel advises the Commissioner on legal matters, is responsible for providing all legal advice to the Commissioner, Deputy Commissioner, and all subordinate organizational components (except OIG) of SSA in connection with the operation and administration of SSA. Responsible for the policy formulation and decision making related to the collection, access, and disclosure of such information in the records of the Social Security Administration; and processing of Freedom of Information requests and appeals (under the Freedom of Information and Privacy Acts).

OFFICE OF THE INSPECTOR GENERAL

The Office of the Inspector General (OIG) is directly responsible for meeting the statutory mission of promoting economy, efficiency and effectiveness in the administration of Social Security Administration (SSA) programs and operations and to prevent and detect fraud, waste, abuse, and mismanagement in such programs and operations. To accomplish this mission, the OIG directs, conducts and supervises a comprehensive program of audits, evaluations and investigations, relating to SSA's programs and operations. OIG also searches for and reports systemic weaknesses in SSA programs and operations, and makes recommendations for needed improvements and corrective actions.

OFFICE OF QUALITY PERFORMANCE

The Office of Quality Performance directs the development of innovative changes to the current Agency quality management program, including the program's initiatives and mechanisms when they are not clearly delineated by statutory authority. Such changes may impact quality management Agency-wide in terms of its programs, policies, and procedures. The Office of Quality Performance is responsible for rendering formal advice and recommendations to Agency executives on a range of issues relating specifically to in-line and end-of-line quality performance management in each of the Agency's core business areas. It works with Deputy Commissioner-level components to direct the Agency-wide quality performance management program, its policies and initiatives involving one or more components of SSA. It also provides oversight for SSA's computer matching operations.

D

Overview of Selected Legislation Pertaining to E-Government

The highest level of federal e-government policymaking is public law. Although some of the public law reviewed below predates the concept of e-government, these pieces of legislation are nonetheless still in place. The overview proceeds generally from broader to more specific information policy and, where possible, links public law to government-wide policy.

THE PAPERWORK REDUCTION ACT OF 1995

The Paperwork Reduction Act (PRA; Public Law 96-511) was originally enacted 1980 as an outgrowth of the 1977 report of the U.S. Commission on Federal Paperwork. The original act recognized the economic cost of the federal government's imposing of paperwork burdens on the public and laid the groundwork for the creation of the Office of Information and Regulatory Affairs (OIRA) in the Office of Management and Budget (OMB) to manage federal information policy centrally. Over time, the PRA has been amended (i.e., in 1986 and 1995) and has resulted in related updates to OMB Circular A-130.[1] It is the PRA that compels federal agencies to get an OMB clearance number on information collections (including such things as forms, surveys, and regulations that require

[1]See Executive Office of the President, Office of Management and Budget (OMB), *OMB Circular No. A-130 Revised*, available at http://www.whitehouse.gov/omb/circulars/a130/a130trans4.html, accessed June 9, 2006. OMB Circular A-130's subject is management of federal information resources.

reporting) and for OMB to aggregate the amount of burden that federal agencies impose on the public annually through the Information Collection Budget. This landmark legislation also had practical implications for agencies, as it forced them to manage information collection as a resource much like they did financial and human resources.

THE INFORMATION TECHNOLOGY REFORM ACT

In many ways, the Information Technology Reform Act (Public Law 104-106; also known as the Clinger-Cohen Act)[2] provided some of the early underpinnings for the e-government movement by requiring agencies to elevate the position of "Senior Official for Information Resources Management" to that of "Chief Information Officer" (CIO). This law recognized the strategic importance of technology in meeting agency objectives, giving the CIO in a federal agency a prominent position that is supposed to report directly to the head of the agency. Consistent with this view, the act created the expectation that agency investments in IT would be evaluated on the basis of the attainment of goals and objectives laid out in agency strategic and tactical plans.

THE PRIVACY ACT OF 1974

Although it is likely the oldest piece of federal information policy that shapes e-government implementation, the Privacy Act of 1974 (Public Law 93-579, as amended)[3] remains quite influential. The Privacy Act is built on the fair information principles outlined by the then-Department of Health, Education and Welfare's policy[4] to provide citizens with insight into their government's stewardship of what the act defines as "sensitive information."[5] Sensitive information includes "information, the loss, misuse, or unauthorized access to or modification of, which could adversely affect the national interest or the conduct of federal programs, or the privacy to which individuals are entitled under . . . the Privacy Act."[6] To

[2]See http://www.cio.gov/Documents/it_management_reform_act_Feb_1996.html, accessed June 20, 2007. See also S.H. Holden and P. Hernon, "An Executive Branch Perspective on Managing Information Resources," pp. 83-104 in P. Hernon, C.R. McClure, and H.C. Relyea, Eds., *Federal Information Policies in the 1990's: Views and Perspectives*, Norwood, N.J.: Ablex Publishing, 1996.

[3]See http://www.usdoj.gov/foia/privstat.htm, accessed June 20, 2007.

[4]See Department of Health, Education, and Welfare Secretary's Advisory Committee on Automated Personal Data Systems, *Records, Computers and the Rights of Citizens*, Cambridge, Mass.: MIT Press, 1973.

[5]P.M. Regan, *Legislating Privacy: Technology, Social Values, and Public Policy*, Chapel Hill, N.C.: University of North Carolina Press, 1995.

[6]See http://www.atis.org/tg2k/_sensitive_information.html, accessed June 20, 2007.

the extent that federal agencies maintain "systems of records" in which it is possible to organize, index, or retrieve the information about a citizen by unique identifier (typically a Social Security number), an agency must post public notice of how the data will be used through a "routine use notice." If any agency wishes to disclose such sensitive data to a third party, the subject of the information must provide consent to such disclosure. For agencies that are receiving or exchanging sensitive information about an individual, which is typical for most e-government transactions, they must take steps to comply with the Privacy Act. Appendix I of OMB Circular A-130 provides the more detailed requirements for agencies to follow when complying with the Privacy Act.

OMB has issued some privacy-related policy pertaining to the use of "cookies" on federal Web sites that grows, in part, out of some of the philosophies of the Privacy Act. OMB issued Memorandum 00-13,[7] which sought to severely limit (and some believe prevent) agencies from using "persistent cookies" that track "the activities of users over time and across different web sites." The memorandum stipulates that federal Web sites should not use persistent cookies unless agencies can meet the following four conditions:

- The site gives clear and conspicuous notice;
- There is a compelling need to gather the data on the site;
- There are appropriate and publicly disclosed privacy safeguards for handling any information derived from the cookies; and
- The agency head gives personal approval for the use.[8]

THE FEDERAL INFORMATION SECURITY MANAGEMENT ACT OF 2002

The Federal Information Security Management Act of 2002 (FISMA; Title III of Public Law 107-347) has superseded the Computer Security Act of 1987. FISMA provides the analytical framework for agencies to assess risk and then to mitigate identified risks for "sensitive information" in federal information systems. In this context, FISMA includes systems that are not national security systems but that contain sensitive information. The following quote from FISMA provides the high-level risk-based security guide for agency decision making in this area. The act states

[7]OMB Memorandum 00-13's subject is privacy policies and data collection on federal Web sites. It is available at http://www.whitehouse.gov/omb/memoranda/m00-13.html, accessed April 14, 2007.

[8]OMB Memorandum 00-13, available at http://www.whitehouse.gov/omb/memoranda/m00-13.html, accessed April 14, 2007.

that agencies shall "identify and provide information security protections commensurate with the risk and magnitude of the harm resulting from the unauthorized access, use, disclosure, disruption, modification, or destruction of . . ." information or information systems maintained by federal agencies or by organizations on behalf of federal agencies. Explicitly in the law, agencies are expected to pursue the goals of the legislation with a cost-effectiveness standard in mind by "implementing policies and procedures to cost-effectively reduce risks to an acceptable level. . . ."[9] Appendix III of OMB Circular A-130 provides the more detailed requirements for agency compliance with FISMA.

THE GOVERNMENT PAPERWORK ELIMINATION ACT OF 1998

The law likely to have provided the most impetus to federal agency e-government efforts was the Government Paperwork Elimination Act of 1998 (Title XVII of Public Law 105-277). It required that individuals or entities that deal with the agencies have the option of submitting information or transacting with the agency electronically by October 2003. It was intended to serve as a transition point from traditional paper-based government and governance at the federal level to the emerging ideals of e-government.[10] Besides setting this target date for agencies to enable electronic information and transactions, it also defined and established the legal sufficiency of electronic signatures necessary for agencies to move away from "wet" signatures on paper. What the law did not do was to specify particular technologies to implement electronic signatures. Instead, the law and subsequent OMB policy[11] was technology neutral, giving agencies wide discretion to match the needs of their technical solutions to the capabilities of their user base and risk mitigation that agency desired.[12]

---

[9]See http://csrc.nist.gov/policies/FISMA-final.pdf, accessed June 20, 2007.

[10]Office of Management and Budget, "Implementation of the Government Paperwork Elimination Act," available at http://www.whitehouse.gov/omb/fedreg/gpea2.html, accessed April 14, 2007.

[11]Office of Management and Budget, Dec. 16, 2003, "E-Authentication Guidance for Federal Agencies," available at http://www.whitehouse.gov/omb/memoranda/fy04/m04-04.pdf, accessed April 14, 2007.

[12]S.H. Holden, "Electronic Authentication Initiatives in the IRS E-File Program: Enabling E-Government Through Electronic Signatures," pp. 984-985 in M. Khosrowpour, Ed., *Issues and Trends of Information Technology Management in Contemporary Organizations*, Hershey, Pa.: Idea Group, 2002.

THE E-GOVERNMENT ACT OF 2002

The E-Government Act of 2002 (Public Law 107-347) and the federal government's strategic plan for e-government provide further impetus for e-government initiatives. The law reaffirmed the federal mandate for e-government by codifying the creation of the associate director for IT and e-government in OMB. It also imposed new annual reporting requirements for OMB to inform Congress of progress toward meeting the goals of the act. Recognizing the need to address public concerns about privacy and e-government adoption, the law required agencies to conduct Privacy Impact Assessments to make privacy considerations more explicit in the development of e-government systems.[13]

[13]S.H. Holden and L.I. Millett, "Authentication, Privacy, and the Federal E-Government," *The Information Society* 21(5):367-377.

E

A Short History of E-Government

Begun approximately a decade ago, e-government refers to the application of the Internet and other information technology (IT) to provide governmental information and services electronically. It offers the potential of increased convenience to the public by making such services available 24 hours a day, 7 days a week, coupled with the advantages of improved accuracy and also reduced cost to the government, deriving from its requiring little or no direct interaction with a government employee. (See Box 1.1 in Chapter 1 of this report for more on the distinction between electronic services and e-government and on the terminology used in this report, generally.) This appendix offers some context for the Social Security Administration's (SSA's) e-government activities and plans. The appendix consists of a brief look at the legal and policy background information as well as history of the federal government's experience with e-government, what the status of e-government is across the United States and to some extent internationally, and then how the SSA's role and progress compare with those of other government agencies.

E-GOVERNMENT DEPLOYMENT IN THE U.S. FEDERAL GOVERNMENT

Until the 1990s the federal government, like much of the business world, used information technology to automate backroom operations, with little emphasis on automating "customer-facing" functions such as information dissemination or service delivery. In many ways, this strategic focus reflected a mainframe processing mentality that had dominated federal IT policy and strategy since the 1960s related to the passage of the Brooks Act of 1949.[1] Under the Brooks Act, one federal agency, the General Services Administration (GSA), was responsible for acquiring IT on behalf of federal agencies.[2] Although the GSA had an elaborate process for delegating this procurement authority to federal agencies, this degree of centralization in IT governance represented a focus on using IT to save money in backroom operations. As a result, the primary criteria for evaluating IT investments were economy and efficiency, so all systems were justified on a "least cost" basis. Interagency IT efforts focused on consolidation efforts such as the Department of Agriculture's National Finance Center for payroll and accounting, which sought to standardize financial systems based on commercial off-the-shelf products and to eliminate duplicative personnel systems.[3]

Starting around the late 1990s, attention began to shift away from simply backroom operations. The federal Chief Information Officer's Council began emphasizing IT projects that offered "service to the citizen." At about the same time, the administration was conducting the National Performance Review (NPR, otherwise known as Reengineering Government) effort, which placed strong emphasis on IT-enabled government. Publication of the NPR report *Access America: Reengineering Through Information Technology* in February 1997 was one of the first occasions on which the federal government began addressing what is now referred to as electronic government.[4]

The projects identified in that report represented a departure from historical emphasis on internal efficiency and economy. The very first initiative involved improving service delivery through technology. This shift from economy and efficiency to service delivery culminated with the first presidential-level directive to federal agencies on e-government

[1]Federal Property and Administrative Services Act of 1949, Public Law 89-306, 40 U.S.C. 759.

[2]The Warner Amendment of 1982 (Public Law 97-86) subsequently exempted certain types of Department of Defense procurements from the Brooks Act and from Section 11 of the Federal Property and Administrative Services Act of 1949. See U.S. Congress, Office of Technology Assessment, *Issue Update on Information Privacy and Security in Network Environments*, OTA-BP-ITC-147, Washington, D.C.: U.S. Government Printing Office, June 1995, p. 106.

[3]S.H. Holden, "The Evolution of Information Technology Management at the Federal Level: Implications for Public Administration," pp. 53-73 in *Public Information Technology: Policy and Management Issues*, G. David Garson, Ed., Hershey, Pa.: Idea Group, 2003.

[4]See http://govinfo.library.unt.edu/npr/library/announc/access/acessrpt.html, accessed June 20, 2007.

in December 1999.[5] The FirstGov.gov Web site, which went online in September 2000, projected the vision that it was "the official U.S. gateway to all government information."[6]

In addition to this federal government activity, other groups were also articulating a vision for e-government. The National Science Foundation (NSF), which adopted the term "digital government" in its efforts, convened a group of prominent researchers in public administration, public policy, information systems, and computer science, along with government practitioners, to articulate a vision for what they called an agenda for digital government research.[7] The Council of Excellence in Government also convened a group of practitioners (from both public and private sectors) and academics to articulate a vision for e-government implementation.[8] The National Research Council, with the support of NSF's Digital Government program, convened a study committee to examine a number of broad technical areas where government investment in IT research would have an impact on the creation of advanced, innovative e-government capabilities.[9]

With the confluence of these visions for how to use IT to improve the delivery of public information and services, it appeared that e-government was coming of age. The public sector saw opportunities for realizing the kinds of gains realized by the private sector's use of e-business, the private sector saw opportunities to sell more products and services, academics saw many research and teaching opportunities, and the public's expectations for how government should work began to evolve as all of these sectors of the economy became increasingly articulate about this new phenomenon. Further presidential highlighting of e-government

---

[5]See history of FirstGov.gov, available at http://firstgov.gov/About.shtml, accessed on June 9, 2006; and William J. Clinton, "Memorandum for the Heads of Executive Departments and Agencies (Electronic Government)," December 17, 1999, available at http://govinfo.library.unt.edu/npr/library/direct/memos/elecgovrnmnt.html, accessed June 20, 2007.

[6]See "About FirstGov.gov," available at http://web.archive.org/web/20060707043959re_/firstgov.gov/About.shtml, accessed August 14, 2007.

[7]Sharon S. Dawes, Peter A. Bloniarz, Kristine L. Kelly, and Patricia D. Fletcher, *Some Assembly Required: Building a Digital Government for the 21st Century*, Center for Technology in Government University at Albany, State University of New York, 1999, available at http://www.ctg.albany.edu/publications/reports/some_assembly/some_assembly.pdf, accessed June 20, 2007.

[8]Council for Excellence in Government, *E-Government: The Next American Revolution*, September 28, 2000, available at http://www.excelgov.org/index.php?keyword=a432c10480be99, accessed June 20, 2007. See also Council for Excellence in Government, *The Blueprint for e-Government*, September 2000 and January 2001, available at http://www.excelgov.org/index.php?keyword=a4338d8c859fc5, accessed June 20, 2007.

[9]National Research Council, *Information Technology Research, Innovation, and E-Government*, Washington, D.C.: National Academy Press, 2000.

institutionalized a strategic emphasis on the application of technology to change the way government works.

With a change in administration in 2001, e-government continued to be a focus in federal agencies and the White House. Two events early in the Bush/Cheney administration cemented the importance of e-government for federal agencies. First, the role of e-government became more prominent in the structure of the Office of Management and Budget (OMB) with the creation of the position of associate director for information technology and e-government, elevating e-government issues from their home at that time under the office responsible for regulatory affairs in OMB. Later in 2001, OMB drafted the first federal strategy for e-government, pulling together 24 "Quicksilver" projects.[10] The federal strategy included four portfolios for the projects: government to citizen, government to business, government to government, and economy and efficiency.

Some specific examples of initiatives from each of the portfolios help describe some of the e-government capabilities available across the federal government. One of the best-known government-to-citizen initiatives is the Internal Revenue Service's (IRS's) Free File initiative,[11] which is a public-and-private partnership to offer free tax preparation and e-filing services to selected taxpayers who have an adjusted gross income of less than $50,000. More broadly, the IRS e-file program has resulted in more than half of all individual tax returns being submitted to the IRS electronically, often times with taxpayers filing their federal and state taxes in one transaction.

For the business-to-government portfolio, the Business Gateway provides a one-stop source of information for businesses seeking to comply with federal regulatory and paperwork burden.[12] MSNBC recognized the Web site for Business Gateway, Business.gov, as its Web site of the week.[13] In the government-to-government portfolio, E-vital is a partnership between the federal government (primarily the SSA) and state governments to share vital statistics, primarily death certificates, electronically.[14] Finally, the federal strategy for e-government also includes a portfolio of initiatives designed to increase the internal efficiency and effectiveness of federal operations. For example, e-payroll is designed to consolidate payroll processes among all federal civilian agencies, thereby

[10]Office of Management and Budget, *E-Government Strategy: Simplified Delivery of Services to Citizens*, Executive Office of the President, Washington, D.C., 2002.

[11]See http://www.whitehouse.gov/omb/egov/c-1-3-IRS.html, accessed June 20, 2007.

[12]See http://www.whitehouse.gov/omb/egov/c-3-5-bg.html, accessed June 20, 2007.

[13]See http://www.msnbc.msn.com/id/14718260/, accessed June 20, 2007.

[14]See http://www.whitehouse.gov/omb/egov/c-2-4-evital.html, accessed June 20, 2007.

simplifying and modernizing the disparate collection of systems that federal agencies now use to pay their employees.[15]

E-government is one of the elements of the President's Management Agenda (PMA), with federal agencies being graded quarterly on achieving the goals of the federal strategy.[16] A presidential memo was issued on the importance of e-government, emphasizing interagency cooperation as a means to provide cost-effective and efficient government services.[17] The inclusion of e-government in the PMA has raised the importance of this initiative in federal strategic management efforts. OMB uses the annual budget process and the "Exhibit 300" Capital Asset and Business Plan Case[18] required as part of annual budget requests from agencies to OMB as a mechanism for the enforcement of agency e-government plans' consistency with the federal e-government strategy and related policies. While e-government was gaining in importance in federal agency plans and administration, the stated goals for federal e-government shifted in 2001 to emphasize data and service integration to support economy and efficiency rather than service delivery.

The e-business experiences of telecommunications and other firms in the private sector indicate that cost reduction and efficiency goals alone are not sufficient—they must be coupled with effectiveness goals that tie the operational effectiveness of their lines of business directly to IT capabilities. However, in business as well as in government, defining and quantifying effectivness measures can be difficult challenges. Nevertheless, government agencies and the publics that they serve can benefit from formulating precise and measurable effectiveness goals.

LEGAL AND POLICY CONTEXT FOR FEDERAL E-GOVERNMENT

The SSA's electronic services plans and initiatives fit within a broader federal context for e-government. The Committee on the Social Security Administration's E-Government Strategy and Planning for the Future acknowledges the complex legislative and statutory environment that all

[15]See http://www.whitehouse.gov/omb/egov/c-4-5-ePay.html, accessed June 20, 2007.

[16]President's Management Agenda (PMA), 2002, available at http://www.whitehouse.gov/omb/budintegration/pma_index.html, accessed June 20, 2007.

[17]George W. Bush, *The Importance of E-Government*, July 10, 2002, available at http://www.whitehouse.gov/omb/egov/g-2-memo.html, accessed June 20, 2007.

[18]The OMB Exhibit 300 Capital Asset and Business Plan Case is used by agencies and by OMB to review the budget justification and business case for major IT investments. See Executive Office of the President, Office of Management and Budget, *OMB Circular No. A-11. Part 7: Planning, Budgeting, Acquisition, and Management of Capital Assets*, June 2005, available at http://www.whitehouse.gov/omb/circulars/a11/current_year/s300.pdf, accessed June 20, 2007.

agencies, including the SSA, must negotiate when looking to broaden or enhance electronic services. This section is a very brief description of some of that context. Like much of federal information law and policy, the legal and policy context for federal e-government across all agencies has been put in place over a number of years through various pieces of legislation and several executive branch initiatives. Appendix D in this report summarizes selected key legislation; although it does not provide an exhaustive review of all the relevant law and policy that might affect the SSA's e-government strategy, it nonetheless provides an overview of legal and policy framework within which the SSA must operate. Rather than having been crafted with e-government in mind, this collection of public law and government-wide policy affecting e-government has accumulated over a period of nearly 25 years. Therefore, federal agencies like the SSA that seek to exploit the benefits of e-government must attempt to do so within a set of legal and policy requirements and constraints that can sometimes be less than clear and consistent. As a general precept, public law is often abstract and somewhat conceptual.

Legislation typically does not provide a statement of requirements that is sufficiently clear and detailed to enable straightforward implementation by agencies. One of the recurring themes in the legislative and policy framework outlined in Appendix D of this report is that agencies have wide latitude on *how* to comply with the *what* of federal information policy. It is also very rare for either public law or government-wide policy to specify particular technologies as part of policy compliance. (Indeed, such specification is suboptimal for myriad reasons.) As a result, the Office of Management and Budget, which has government-wide policymaking and oversight responsibility, typically interprets the law through its OMB Circular A-130, *Management of Federal Information Resources*.[19] Other federal agencies, including the General Services Administration (GSA) and the National Institute of Standards and Technology (NIST) also have government-wide information policy responsibilities. Federal agencies then often further interpret government-wide information policy to fit their organizational context. In multibureau federal departments, each bureau may even localize the policy further.[20]

---

[19]See Executive Office of the President, Office of Management and Budget (OMB), *OMB Circular No. A-130 Revised*, available at http://www.whitehouse.gov/omb/circulars/a130/a130trans4.html, accessed June 9, 2006.

[20]Stephen H. Holden and Peter Hernon, "An Executive Branch Perspective on Managing Information Resources," pp. 83-104 in Peter Hernon, Charles R. McClure, and Harold C. Relyea, Eds., *Federal Information Policies in the 1990's: Views and Perspectives*, Norwood, N.J.: Ablex Publishing, 1996.

STATE OF THE PRACTICE IN E-GOVERNMENT

The U.S. federal government is not alone in realizing the potential benefits of e-government. E-government services are also offered by all of the states, many municipalities, and indeed by the governments of most of the world's developed countries. A number of efforts by varied groups, including the United Nations (UN) and the American Society for Public Administration (ASPA), researchers at Brown University, and various consultancies have begun to compare e-government offerings internationally. Typically, these groups have found that the quality of the e-government offerings of the U.S. federal government is among the best worldwide, although public adoption of such services has been found to be higher in some other countries.[21] For example, Accenture's January 2005 survey of people in the United States found that over 55 percent had made at least some use of e-government; the same survey found that almost 70 percent of those surveyed used the Internet at least once a month. Accenture also found that reported use of e-government was higher in Australia, Canada, and Singapore (with reported Internet usage levels similar to those in the United States), and in Norway, Denmark, Sweden, and Finland (with higher percentages of people using the Internet at least monthly).[22]

A study in 2004 conducted by the UN and ASPA found that the United States offers the most sophisticated e-government services of all the UN member countries studied.[23] However, the U.S. public still makes only a rather limited use of e-government services, restricting their utilization primarily to electronic access to government information and publications. The great majority of the public's interactions with its government is still either on the phone or in person. Significantly, at the same time, the public's use of the Internet and electronic services for banks and other financial institutions, as well as for retail purposes, is far more substantive, generally centering on actual financial transactions, in addition to obtaining information. One possible reason for this difference in the nature of the public's interactions may be that current e-government services beyond provision of information do not meet users' needs as effectively as the services offered by the private sector (see Chapter 2 in

[21]A contributing factor toward greater usage in some other countries may be differing attitudes toward privacy and how personal information is handled by governments and the private sector.

[22]Accenture, 2005 E-Government Report, "Leadership in Customer Service: New Expectations, New Experiences," available at http://www.accenture.com/xdoc/ca/locations/canada/insights/studies/leadership_cust.pdf, accessed June 20, 2007.

[23]United Nations, *Global E-Government Readiness Report 2004: Towards Access For Opportunity*, New York: United Nations, 2004, available at http://www.unpan.org/egovernment4.asp, accessed June 20, 2007.

this report for more on this topic).[24] For instance, some e-government services require users to supplement electronic transactions with paper signature documents or do not offer the complete set of electronic transactions that users might like to use. For example, in some places it is possible to apply for a duplicate birth certificate online, but in other places it is not. In other cases, users still have to understand the interworkings of government agencies to know where to look to retrieve information electronically or to complete transactions. The success of the private sector in meeting public needs with electronic services surely heightens public expectations of government agencies such as the SSA. Increasing demands from users beyond individual beneficiaries, such as the states and other federal agencies, will increase the pressure on the SSA, as will increasing the pressures for cost reductions and efficiency improvements. Taken together, it seems reasonable that all of these pressures will inevitably cause the SSA to move toward increases and improvements in its electronic service provision.

Empirical Studies

The SSA and other governmental agencies have begun to respond to these pressures, and their responses have been growing in both quantity and quality. Internal federal government evaluations, such as the quarterly PMA (described above), have both tracked and precipitated these improvements. Other external evaluations have done so as well. (Chapter 4 in this report describes some empirical results on user attitudes and behavior generally.) The longest-running set of external evaluation studies of e-government deployment results have been published by Darrell West at Brown University's Taubman Center for Public Policy.[25] Dating back to 2001, these studies have evaluated e-government offerings, primarily based on analysis of government Web site content. These analyses have compared Web offerings internationally (among national governments), nationally in the United States (across federal agencies), and also

[24]Council for Excellence in Government, *The New E-Government Equation: Ease, Engagement, Privacy and Protection*, 2003, available at http://www.excelgov.org/usermedia/images/uploads/PDFs/egovpoll2003.pdf, accessed June 20, 2007; and Elena Larsen and Lee Rainie, *The Rise of the E-Citizen: How People Use Government Agencies' Web Sites*, 2002, available at http://www.pewinternet.org/reports/toc.asp?Report=57, accessed June 20, 2007.

[25]See Darrell M. West, *State and Federal E-Government in the United States*, Providence, R.I.: Taubman Center for Public Policy, 2005, available at http://www.insidepolitics.org/egovt05us.pdf, accessed June 20, 2007.

among the various U.S. states. There have been studies that examined and compared social security administrations internationally as well.[26]

Those studies that looked specifically at federal agencies gave positive weight to features such as online publications, online databases, and the availability of online transactions. They typically also examined features such as disability access, privacy policy, security policy, Web site personalization, personal digital assistant (PDA) accessibility, and readability level. Features such as advertisements, premium fees, and user payments or fees were considered negative and detracted from an agency's e-government score. Such studies have typically given the SSA's e-government efforts very favorable evaluations compared with its public-sector peers.[27] West also finds that federal agencies continue to struggle with issues like broken links, poor compliance with accessibility standards, and readability levels that exceed the capabilities of average users. In general, federal agencies are putting more transactions online and are increasingly posting privacy and security policies on their Web sites.[28] West's findings seem to be confirmed by data obtained from the University of Michigan's American Customer Satisfaction Index (ACSI)—see below.

West's findings suggest that agencies such as the SSA have much work to do. The agencies have steadily been taking steps to address shortcomings in their online presences. More effective grappling with broader issues is still needed, however, and a high score in the limited West evaluation scheme can certainly not be taken as a cause for complacency. In the

---

[26]A recent study examined the impact of automation on social security administrations across 13 countries. While all 13 would likely be considered developed countries, some countries had different philosophies on their approach to social services provision (i.e., whether they were Scandinavian "social democratic," European "corporatist," or Anglo-American "liberal" countries). The significance of this would be somewhat varied service availability online depending on the regime values of the country's social security system under study. The historical perspective provided by the study is helpful in explaining that automation really came to these national-level social security organizations in the mid-1950s, with the U.S. SSA beginning around 1955, third on the list of countries studied. The report examined the goals for these automation efforts between 1985 and 2000 and found, not too surprisingly, that the primary aims initially were very focused on cost cutting and productivity improvement. Between 2000 and 2004, though, the list of goals grew considerably to include improving information for both in-house staff and outside users, preventing and reducing program fraud, and responding to user demand for online information and services. A head-to-head comparison of the level of aggregate online functionality found the United States with the highest ranking—17 out of a possible 28 points—and the Netherlands coming in lowest with a score of 10. See Michael Adler and Paul Henman, with Jackie Gulland and Sharon Gaby, *Computerisation and E-Government in Social Security: A Comparative International Study*, Arlington, Va.: IBM Center for the Business of Government, 2005, available at http://www.businessofgovernment.org/pdfs/AdlerReport.pdf, accessed June 20, 2007.

[27]Darrell M. West, *State and Federal E-Government in the United States*, 2005.

[28]Darrell M. West, *State and Federal E-Government in the United States*, 2005.

committee's view, findings such as these suggest that the efforts of agencies such as the SSA that are seeking a forward-looking strategy in the electronic services arena should be compared not with other government agencies, but rather with those of the private-sector financial institutions that are shaping the expectations of the public, especially in regard to service levels expected from large financial institutions (as described in Chapter 2 of this report). While this comparison fails in some ways (e.g., aggressiveness and the role of competition, funding sources, client base, and frequency of interaction), there are useful lessons and insights to be gleaned.

American Customer Satisfaction Index

Since 1994, the University of Michigan has published a series of snapshots of customer impressions of a variety of services in its American Customer Satisfaction Index (ACSI).[29] The ACSI provides government agencies (and commercial ventures) with an independent measure of consumer experience. The index relies on a model that includes data on customer expectations, perceived quality, perceived value of information, and on customer complaints and customer loyalty.[30]

The ACSI started reporting separately on federal agency performance in 1999, about the same time that federal agencies started enabling and promoting their e-government offerings. The ACSI recently added an evaluation for e-government services among federal agencies. Not too surprisingly, the ACSI scores for federal government agencies have been going up gradually since the e-government index was created in late 2003, with the aggregate satisfaction score in September 2005 being 73.5, against an 80.0 score considered exceptional for online transactions either in the public or private sector.[31]

In the September 2005 reporting of the ACSI e-government satisfaction index, the SSA did relatively well, ranking first of eight federal agency sites in the e-commerce/transaction category (with a score of 87)

---

[29]See "About ACSI," available at http://www.theacsi.org/index.php?option=com_content&task=view&id=49&Itemid=28, accessed June 19, 2007.

[30]"ACSI reports scores on a 0-100 scale at the national level and produces indexes for 10 economic sectors, 43 industries (including e-commerce and e-business) and more than 200 companies and federal or local government agencies. In addition to the company-level satisfaction scores, ACSI produces scores for the causes and consequences of customer satisfaction and their relationships." See "About ACSI," available at http://www.theacsi.org/index.php?option=com_content&task=view&id=49&Itemid=28, accessed June 19, 2007.

[31]ForeSee Results, "American Customer Satisfaction Index, E-Government Satisfaction Index," Ann Arbor, Mich., December 15, 2005. Archived ACSI scores and commentaries are available at http://www.theacsi.org/index.php?option=com_content&task=view&id=68&Itemid=57, accessed June 20, 2007.

for the Web site "Internet Social Security Benefits Application." Of 44 government Web sites rated in the news/information category, the SSA was ranked first (with a score of 91) for the Web site "Help with Medicare Prescription Drug Costs" and seventh (with a score of 81) for the Web site "Social Security Business Services Online." The SSA's "Frequently Asked Questions" customer-help page (with a score of 75) and its information for disability benefits (with a score of 71) ranked in the middle and closer to the bottom, respectively, of the Web sites studied. In the March 2006 report, the SSA had the top 4 (of 11) sites in the e-commerce/transaction category and the 2 top ratings in that category (with scores of 86), for "Help with Medicare Prescription Drug Plan Costs" and "Internet Social Security Benefits Application." However, the SSA's main Web site (www.socialsecurity.gov) ranked toward the bottom of the agencies' sites studied, with a satisfaction score of 70 (the top-ranked site scored 82).[32]

There has been a similarly favorable response to the availability of online e-government transactions. A 2005 ACSI study of the IRS compared the public's satisfaction with paper filing of tax returns to the public's satisfaction with electronic filing.[33] The ACSI score for paper filing was 50, while satisfaction with electronic filing was 77. While there are distinctions between the kinds of services that the SSA offers and its client base as compared with the services and client base of financial institutions, increased use of online financial services in the private sector may shape expectations, at least on the part of some, for a similarly broad availability of online transactions from agencies such as the SSA that provide important financial services.

[32]Scores reported by ACSI can be found at http://www.theacsi.org/index.php?option=com_content&task=view&id=27&Itemid=62, accessed June 20, 2007.

[33]The IRS worked with third-party vendors to implement electronic filing (see Chapter 4 in this report).